T0136342

Forecasts and
Environmental Decisionmaking

Social Impact Assessment Series
C. P. Wolf, General Editor

Women and the Social Costs of Economic Development: Two Colorado Case Studies, Elizabeth Moen, Elise Boulding, Jane Lillydahl, and Risa Palm

Social Impact Assessment and Monitoring: A Cross-Disciplinary Guide to the Literature, Michael J. Carley and Eduardo Bustelo

Integrated Impact Assessment, edited by Frederick A. Rossini and Alan L. Porter

Public Involvement and Social Impact Assessment, Gregory A. Daneke, Margot W. Garcia, and Jerome Delli Priscolli

Applied Social Science for Environmental Planning, edited by William Millsap

Guide to Social Impact Assessment: A Framework for Assessing Social Change, Kristi Branch, Douglas A. Hooper, James Thompson, and James Creighton

Social Impact Analysis and Development Planning in the Third World, edited by William Derman and Scott Whiteford

Differential Social Impacts of Rural Resource Development, edited by Pamela D. Elkind-Savatsky

Forecasts and Environmental Decisionmaking: The Content and Predictive Accuracy of Environmental Impact Statements, Paul J. Culhane, H. Paul Friesema, and Janice A. Beecher

A Systems Approach to Social Impact Assessment: Two Alaskan Case Studies, Lawrence A. Palinkas, Bruce Murray Harris, and John S. Petterson

Forecasts and Environmental Decisionmaking

The Content and Predictive Accuracy of Environmental Impact Statements

Paul J. Culhane,
H. Paul Friesema,
and Janice A. Beecher

Routledge
Taylor & Francis Group

LONDON AND NEW YORK

First published 1987 by Westview Press, Inc.

Published 2018 by Routledge
52 Vanderbilt Avenue, New York, NY 10017
2 Park Square, Milton Park, Abingdon, Oxon OX14 4RN

Routledge is an imprint of the Taylor & Francis Group, an informa business

Library of Congress Catalog Card Number: 87-50986
ISBN 13: 978-0-367-00597-9 (hbk)
ISBN 13: 978-0-367-15584-1 (pbk)

Contents

Tables and Figures

TABLES

FIGURES

Preface

This research was supported by grant PRA-8119299 from the National Science Foundation, Division of Policy Research and Analysis, to Northwestern University. Its purpose was to evaluate the predictive accuracy of the forecasts in a sample of federal environmental impact statements (EISs). Since federal agencies devote much staff time to writing environmental assessments, the accuracy of those assessments is an intrinsically important issue. Indeed, the accuracy of environmental assessments has become an international question, as indicated by a 1985 conference in Banff, Canada, a 1986 special issue of <u>Environmental Monitoring and Assessment</u>, and other projects devoted to the subject.

In this book we evaluate the match between the reality of EIS writing and the ideals of rational analysis, the pathologies of project implementation, and other issues that extend beyond the predictive accuracy of EISs. We examine, in effect, a major federal attempt to impose rationalistic reforms on government decision makers.

This study has roots in several institutions. The senior authors began examining the NEPA process in 1973 at Northwestern University's Center for Urban Affairs, and planned to evaluate the predictive accuracy of EISs as part of their program of NEPA studies (Culhane and Friesema 1978: 30-32). We developed specific plans for the study as part of the environmental policy program at The Institute of Ecology during 1980-1981, and began the study after that program was moved to Northwestern University. TIE had sponsored an EIS monitoring project during the 1970s (cf. Winder and Allen 1975, Cromwell et al. 1977), and we were pleased to continue that work.

We are responsible for the work reported in this book, of course, even if we acknowledge its insights more readily than its errors. Paul Friesema is the principal author of Chapter 2, with some contributions by Paul Culhane; Jan Beecher is the author of Chapter 3; and Paul Culhane is the author of the remaining chapters. The conclusions are ours and do not reflect the views of the National Science Foundation or Northwestern University.

Nonetheless, we wish to gratefully acknowledge the help we received on this project from Charles Matzke and Michael Peddle, who assisted us during our 20,000 miles of fieldwork, and Sarah-Katherine McDonald and Charles LeHew, who assisted in the postfieldwork phase of the project. We also thank Barbara Angelescu and Gaye Haverkos of Northwestern's Center for Urban Affairs and Policy Research for their capable staff assistance.

We have benefited throughout this project from the encouragement and wisdom that we received from our advisory committee: Dinah Bear, Council on Environmental Quality; Bruce Blanchard, Department of the Interior; Paul Brace, Department of Housing and Urban Development; Gordon Enk, International Paper Company; Allen Hirsch, Dynamac Corporation; Richard Liroff, Conservation Foundation; Orie Loucks, Holcomb Research Institute; Richard McCleary, University of New Mexico; Richard Morrison, National Science Foundation; and Robert Stern, New Jersey Department of Environmental Protection. The advisory committee and C. P. Wolf, general editor of Westview's social impact assessment series, also provided invaluable comments on our draft manuscript.

We also thank Diane Culhane for her usual fine editing, and Barbara Ellington and Bruce Kellison of Westview Press for their patience despite numerous missed deadlines.

Finally, this book is dedicated to Kimberly Lally, who helped delay completion of the book but enriched the senior author's life in the process.

Paul J. Culhane
H. Paul Friesema
Janice A. Beecher

1
Environmental Impact Statements and Federal Decisionmaking

The National Environmental Policy Act fundamentally reformed federal resources decisionmaking. During the first decade following NEPA's January 1970 signing, federal bureaucrats wrote 10,475 environmental impact statements. These documents ranged from early, skimpy reports of under a dozen pages to the massive 58-volume 1981 final EIS on Nevada-Utah basing of the MX missile.[1] Environmental impact statements have seemed such a good idea that the act's procedures have been copied, with greater or lesser faithfulness, by half the state governments in the U.S. (Hart and Enk 1980), the governments of Canada and most European community nations (Wandesforde-Smith 1978), and federal agencies concerned with "inflationary impacts," "arms control impacts," "regulatory impacts," and "urban impacts." Former Chief Justice Warren Burger, of the federal judiciary that ensured EISs would have a major impact on the federal bureaucracy, even called for "judicial impact statements" on the effects of new legislation on the courts' workload. Clearly, the concept of environmental impact statements is admired and imitated. A separate question is, how good are the documents themselves?

Academic observers and participants in federal policymaking usually agree that the EIS process has been a beneficial reform. However, they disagree about how this federal assault on the pulpwood forests of North America has improved federal decisionmaking. Depending on whom one chooses to believe, NEPA has brought the technical precision of science to bear on resource decisionmaking, added environmentally sensitive officials to previously insensitive bureaus' staffs, or opened up otherwise parochial agency decision processes to public scrutiny.

1

Naturally, an operating definition of a "good" EIS depends upon which view of the NEPA process one focuses on.

This book examines the first view of NEPA reforms. If EISs are rational, analytical documents, they should contain comprehensive, competent predictions about the consequences of agencies' proposed actions. Several criteria are embodied in the phrase "comprehensive, competent predictions." First, the predictions should meet some accepted standard of technical sophistication. Different advocates advance slightly different standards, such as the use of quantified forecasting models or reliance on state-of-the-art scientific theory about a given type of impact. But rationalist reformers generally presume that the more analytical, scientific, and quantified an EIS prediction is, the better. Second, the predictions should be correct. From a rationalist point of view, if a collection of technical, quantified, and seemingly sophisticated predictions is systematically wrong, then any decision based on those predictions can be right only by virtue of dumb luck.

PUBLIC DECISION THEORY

The rationalist theory of decisionmaking has subtly influenced prescriptive theories of environmental assessment. The classic rational model first entered the literature on public administrative behavior in Herbert Simon's celebrated treatise <u>Administrative Behavior</u>. Simon (1947: 67) described fully rational decisionmaking as a four-step process:

1. Decisionmakers are assumed to agree on the goals that govern a given decision;
2. Decisionmakers identify all alternative courses of action that are relevant to their goals;
3. Decisionmakers identify all relevant consequences of each alternative; and
4. Using some appropriate calculus, decisionmakers compare the sets of consequences and decide upon the optimum alternative.

Simon credited the mathematical economists von Neuman and Morgenstern (1944) for influencing his formulation of the classic rational model. Even though the four steps are nowhere to be found in their book, Simon's citation is apt since the economists' "rational economic man" knows his "preference function" (goals), has "perfect information"

2

(i.e., about alternative actions and their consequences), and chooses the action with the highest "net benefit," that is, the one with the best mix of goal accomplishment and side effects.

The practicality and desirability of rational decision-making in bureaucracies have been debated within the fields of public policy and organization theory since Simon first described the model. Decision theory can be subdivided into various approaches. However, to appreciate decision theory's relevance to NEPA reforms, it is easiest to divide decision theorists into the camps of the rationalists, generally led by economists, and the antirationalists, led by political scientists and psychologists.

Antirationalist Decision Models

In Simon's original contributions to decision theory, the rational model was set up as a "strawman" that the author picked apart before describing his own theory of decisionmaking. (The same holds true for the contributions of other leading theorists like Lindblom and Allison.) Simon argued that human decisionmakers generally cannot meet the information-processing demands of the rational model. Moreover, comprehensiveness is not cost-effective in normal organizational decisionmaking since the marginal utility of the "best" decision is usually less than the marginal cost of a comprehensive search for it. Using basic principles of cognitive psychology, Simon advocated a "satisficing" model (March and Simon 1956) in which decisionmakers focus on selected aspects of a decision problem and then adopt the first satisfactory solution that they find during a search through alternatives. The products of de novo satisficing searches are then added to the organization's repertory of standard operating procedures. The starting point in organizational decisionmakers' searches for solutions is existing organizational standard operating procedures, or "preformed decisions" (Kaufman 1960).

Charles Lindblom (1959) concurred with Simon's judgment that comprehensive analysis is rarely feasible in public bureaucracies. Lindblom's distinctive observation is that participants in public policymaking routinely disagree on the proper goals to be optimized by a decision. That is, the contestants in a political decision process have differing interests or philosophies that lie at the root of their disagreements over a decision. Thus, decisionmakers must try to arrive at a consensus that different partici-

3

pants can agree on for different reasons, a process Lindblom (1965) describes as "partisan mutual adjustment." By extension, present decisions tend to differ little--or "incrementally"--from past decisions, since yesterday's decisions represented a satisfactory consensus among roughly the same set of actors as participate in today's decisions.

Most work in decision theory since the 1950s has replowed the ground first turned by Simon and Lindblom. This field cannot be concisely summarized in a few pages, but two prominent works demonstrate the pervasive influence of Simon's and Lindblom's original insights.[2] First, Aaron Wildavsky's classic description of federal budgeting, Politics of the Budgetary Process (1964: Ch. 2), combines Simon's and Lindblom's concepts: budgeting is "simplified" (i.e., not comprehensively rational), Wildavsky argues, and budgetmakers "satisfice" by bargaining and role-playing among themselves, resulting in budgets that change only incrementally. Second, in the foreign policy literature, Graham Allison's (1970) Essence of Decision urges readers to consider two models of decisionmaking as alternatives to the version of the rational actor theory then in vogue among foreign-policy analysts. Allison's "organizational process model," which was explicitly derived from the theories of Simon and his colleagues, pictures decisions as products of organizations acting according to standard operating procedures. Allison's "governmental politics model," which was explicitly derived from the theories of Simon and his colleagues, develops Lindblom's partisan mutual adjustment theory into a depiction of decisions as results of negotiating and influence games among governmental actors who have different interests and perceptions of the "faces" of issues.

These models are all cut from the same cloth. They all involve critiques of the classic rational-comprehensive decision model, critiques based on empirical arguments about the cognitive psychology of decisionmakers. All are arational decision theories. They do not portray decisionmaking as irrational, unreasonable, illogical, or inefficient. The arational models describe decisionmakers whose cognitive psychology differs fundamentally from that of a rational-comprehensive analyst. Arational decisionmaking, however, is usually more cost-effective (March and Simon 1956, Olson 1984) and politically sensible (Lindblom 1959, Wildavsky 1964) than technically "rational" decisionmaking.

Systems Analysis

Rational-comprehensive decision theory did not enjoy the status of conventional wisdom during or before the 1950s. Rather, early decision theorists used it as a counterpoint to their descriptions of real-world decisionmaking. However, by promising the "best" decision, rationalism remained quite attractive as a normative and prescriptive model of decisionmaking. Bona fide advocates of rational decisionmaking thus entered the public administration literature in the 1960s--years after Simon and Lindblom had picked apart the classic rational model. These true advocates of rationalism generally operated under the label of "systems analysts."

Systems analysis originated during World War II as a set of mathematical or "operations research" methods for solving complex Allied strategic and logistical problems. Edward Quade (1966), of the Rand Corporation, has been among the leading proponents of systems analysis in decisionmaking. He depicts systems analysis as a cycle in which decision-makers and their analysts (1) define the problem to be solved and select their objectives, (2) design alternative programs, (3) collect data and build models to estimate each alternative's consequences, (4) calculate the cost and effectiveness of each alternative, (5) calculate the sensitivity of these cost-effectiveness results to data uncertainties and changed assumptions, and (6) question assumptions, reexamine objectives, investigate new alternatives, and repeat the analysis.

At the core of systems analysis one finds the standard elements of the classic rational-comprehensive model: goal selection, identification of alternatives and their conse- quences, and an optimizing decision. However, Quade adds two important elements to the simple rational model. First, he expects analysts to use technically informed models to estimate policy impacts and cost-benefit analysis to weigh efficiency. While nonquantitative techniques could be appropriate if used by analysts in an objective fashion, he prefers mathematical models because quantification suppos- edly forces analysts to be clear and exposes biases. Second, systems analysis is a cyclic process. Quade depicts the steps in systems analysis by a circular series of clock- wise arrows. This figure does not include a step in which an optimum choice is made, suggesting the eternal triumph of analysis over decision. It does, however, imply that a decision will be reached and implemented at step 4.5 in some analysis, and that steps 5 and 6 will involve policy evalua-

5

tion. (See also Brewer and deLeon 1983.)

Quade's exposition of systems analysis as a decision model was prepared as a staff paper for Rand's project on planning-programming-budgeting (Novick 1965). PPB was devised to reform the incremental federal budgeting practices Wildavsky described. Essentially, all activities with the same basic goal were to be grouped in a budget program; systems analysis models would be used to estimate outputs, alternatives, and cost-effectiveness; and budgets would be efficiently planned over multiyear budget cycles. After some apparent success during Secretary Robert McNamara's crusade to subdue interservice rivalry in the Department of Defense, President Johnson ordered PPB implemented throughout the federal government. Thus, in the late 1960s PPB seemed to be a triumph for advocates of systems analysis.

Just as PPB was intended to reform incremental budgeting, so Quade's systems analysis aimed to reform arational decisionmaking. Arational decision models are based on the premise that human decisionmakers cannot process the data needed to calculate comprehensive, optimizing decisions. Systems analysts believed that emerging decisionmaking tools such as linear programming, aided by then-new computer technologies, removed the cognitive barriers to more rational choice (cf. Simon 1960). Quade's systems analysis was also a form of "rationalist incrementalism." His decision cycle implied that future decisions ought to be refinements of past decisions. But the decisionmaker in such a cyclical process would be a systematic, analytical optimizer—not someone who "muddled through."

THE NATIONAL ENVIRONMENTAL POLICY ACT

The NEPA process is partly a product, like PPB and systems analysis, of the rationalist decision reform movement of the 1960s. NEPA's advocates included critics of federal incremental policymaking. In the years before NEPA, agencies approved projects by conducting a satisficing comparison of a project's objectives with the agencies' narrow mission and tended to ignore many potential adverse consequences (Caldwell 1982: 75-77). For example, their assessment might be limited to a simple engineering feasibility study, perhaps accompanied by a narrow cost-benefit analysis. In the view of environmentalist critics, such behavior involved a narrow pursuit of economic development objectives—too often objectives with dubious economic credentials—and a disregard for adverse environmental consequences. Congress thus found itself presented with embarrassing unanticipated consequences and "environmental disasters." Moreover, members of the authorizing committees of Congress were often frustrated by the fact that satisficing agencies invariably presented them with a single administration-endorsed proposal and no alternatives. Congress thus felt itself put in the position of taking or leaving a complex, technical resource project that most senators and representatives were ill equipped to evaluate.

The National Environmental Policy Act was passed by Congress in December 1969, after less than a year of relatively quiet legislative deliberation. President Nixon signed NEPA on January 1, 1970, hailing it as the first act of the "environmental decade." The short statute begins with the declaration that the federal government ought to "maintain conditions under which man and nature can exist in productive harmony" and ends with the establishment of a small White House staff agency.[3] NEPA's key provision (at least in retrospect) prescribes the rational-comprehensive decision model as a cure for the incrementalist maladies afflicting federal resources policy. Section 102(2)(C) orders agency officials to write environmental impact statements evaluating all reasonable alternatives to a proposed project and analyze all relevant consequences of each alternative:

[A]ll agencies of the Federal Government shall . . . include in every recommendation or report on proposals for legislation and other major Federal actions significantly affecting the quality of the human environment, a detailed statement by the responsible official on—
(i) the environmental impact of the proposed action, . . .
(iii) alternatives to the proposed action. . . .

NEPA's plain statutory language does not explicitly require strict adherence to the full-blown rational-comprehensive model. The statute does not require a lead agency to identify all alternatives, just "alternatives to the proposed action." It does not require consideration of all consequences, only "the environmental impacts of the proposed action." However, federal court precedents soon required agencies to consider all reasonable alternatives. The courts held that EISs are only required in cases of significant impacts on the natural environment, but once a case passes that threshold, EIS writers must consider the full range of significant impacts--social and economic as well as biological and physiographic.

Nonetheless, one can easily read into NEPA a mandate for the kind of technically sophisticated analysis envisioned by Quade and other rationalist reformers. For example, the subsections in section 102(2) immediately preceding the EIS mandate require all federal agencies to

(a) Utilize a systematic interdisciplinary approach which will insure the integrated use of the natural and social sciences and the environmental design arts in . . . decisionmaking which may have an impact on man's environment;
(b) Identify and develop methods and procedures . . . which will ensure that presently unquantified environmental amenities and values may be given appropriate consideration in decisionmaking along with economic and technical considerations.

Key words like "systematic," "sciences," "methods," and "quantified" suggest systems analysis procedures. Inter-disciplinarity, another concept in vogue in the 1960s, is also a rationalist reform. The core officials in many federal resource agencies, so the argument goes, tradi-tionally come from a single profession or discipline, such as civil engineers in the Corps of Engineers or foresters in the Forest Service. As Commoner (1971) argues in The Closing Circle, a bible of environmental activists, professionals tend to focus on the narrow range of causes and effects that are within the purview of their discipline. Involving professionals from a range of disciplines in decisionmaking would facilitate the identification of all possible consequences.

In the most expansive vision of NEPA as a decision reform, Lynton Caldwell (1982: 2) argues that the statute recruited science to redirect policymaking:

Enlistment of science on behalf of policy was necessary because only through science, broadly defined, could the impact of man's activities upon the environment adequately be assessed and remedial measures be applied where needed. . . . To achieve NEPA goals, an inte-grated interdisciplinary use of science was necessary to address complex and interrelated environmental problems. Recognition of the need to redeploy and reintegrate scientific knowledge to respond to the complex

8

challenges of environmental policy gave practical expression to the theoretical unity of science.

Caldwell expects, at a minimum, that environmental assessments will be informed by state-of-the-art science. Caldwell served as a consultant to the Senate in the drafting of the act and played such a key role in the Senate hearings that he is widely regarded as the academic godfather of NEPA. Thus, read together with the statute's references to the "use of . . . sciences" and periodic attempts to improve the scientific and technical quality of EISs (e.g., Enk and Hart 1980), Caldwell's argument carries considerable authority.

Caldwell also suggests that NEPA could further the advancement of science itself. The great scientific vision of the 1960s, general systems theory (which is distinguishable from systems analysis), sought a holistic science that could transcend the narrow reductionism of individual disciplines. Ecological biologists (who have long differed with their reductionist confreres, the microbiologists) have significantly influenced developments in both general systems theory and environmentalism. Ecological biologists' and general systems theorists' insights are quite useful in understanding environmental problems, and NEPA's supporters generally subscribe to a thoroughly holistic view of environmental assessment (e.g., Cromwell et al. 1977). Nonetheless, the politicians in Congress undoubtedly did not intended that NEPA bring about the epistemological ascendancy of general systems theory.

Liroff (1976: 16-18) argues that NEPA's drafters, influenced by Professor Caldwell's testimony, intended EISs to contain the same systematic analyses as PPB budget documents. However, NEPA did not require the final optimization step in the rational-comprehensive decision sequence. Optimization can only be understood in terms of the first step in the rationalist sequence, goal agreement. Section 101's goals are so vague and global as to be of minimal use in making choices about individual projects. NEPA also did not place environmental values above all others. Section 105, following one of the few disputes in the act's legislative history, disclaimed any inference that the act repealed existing agency missions on environmental grounds. NEPA, in short, did not provide federal decisionmakers with what decision theorists would call a preference function.

Subsequent legal interpretations of the statute did little to clarify the "substantive" mandate of NEPA, that

9

is, the relative priority decisionmakers should give to environmental and nonenvironmental values. The courts, led by the D.C. Circuit's Calvert Cliffs (1971) decision, held that EIS writers must "balance" environmental with technical and economic considerations. But because of the law on judicial review under the Administrative Procedures Act, the courts have invariably held EIS writers to a procedural standard of full consideration of alternatives and their impacts. The Council on Envionmental Quality's regulations (CEQ 1978) governing the NEPA process, as another example, required agencies to identify an "environmentally preferred alternative," but following a policy debate on this point, decisionmakers were not required to choose that alternative. In other words, NEPA mandated a truncated rational decision-making process. EIS writers must identify all alternatives and consider all consequences in a technically comprehensive manner. However, NEPA did not establish an order for goals in the process or require them to reach an optimal decision. NEPA led the horses to the waters of rational-optimizing decisionmaking, but it did not require them to drink.

IMPACT ASSESSMENT METHODS

Whether or not the act required fully optimal decisions, most authors in the prescriptive literature on environmental assessment have accepted NEPA's invitation to engage in rational-technical analysis. During the first few years following the passage of NEPA, a number of frameworks were proposed in response to the instant demand NEPA created for environmental assessment methodologies. The frameworks that appeared within three years of NEPA's passage were, for the most part, produced by pollution-control engineers, water resources planners, and some biological ecologists whose experience was based on their work in water project and pollution control planning during the pre-1970 formative years of environmental management (cf. Warner and Preston 1973). The prescriptive literature also contains a variety of practitioner guidebooks on the EIS process.[4]

The distinctions among the many environmental assessment frameworks are less relevant than the core of rational-technical analysis that all espouse in one way or another. Therefore, we shall use one of the best expositions of EIS methods to illustrate a state-of-the-art assessment framework at the time NEPA became firmly institutionalized in federal resources policymaking. Larry Canter's Environmental Impact Assessment (1977: 20) adopts the science-driven view of NEPA espoused by Caldwell:

Assessments of the environmental impacts of proposed actions require systematic, reproducible, and inter-disciplinary approaches. Systematic denotes an all-inclusive, orderly, and scientific consideration of potential impacts on the physical, biological, cultural, and socioeconomic aspects of the environment. . . . Finally, inputs are required from many disciplines to ensure that a complete analysis has been accomplished.

Canter's assessment framework can be inferred from the "basic steps" outlined at the beginning of his six chapters on particular types of impacts, plus his general methods chapter (Ch. 4-10). It consists of nine basic steps for impact prediction and assessment:

1. Identify the byproducts of a project that might affect some environmental system.
2. Inventory the existing special problems, historical trends, or conditions of that system that a project could affect.
3. Identify characteristics of a site or area that could interact with a project's byproducts, and inventory any existing load of the byproduct within the project area.
4. Learn the environmental standards that apply to the impact, if any.
5. Using appropriate methods, calculate the "mesoscale" (i.e., regional) impacts of each alternative.
6. Using appropriate methods, calculate the "micro-scale" (i.e., local) impacts of each alternative.
7. Consider construction-phase impacts, especially as they differ from long-term or post-construction impacts.
8. If an impact is worse than applicable standards allow, evaluate mitigation or control measures.
9. Compare these impacts with other types of impacts, using some uniform scaling method, before selecting the preferred alternative.

Canter varies his framework somewhat in his substantive chapters. Instead of steps 1-8 above, he presents 12 steps in the chapter on water impacts, nine steps in the chapter on air impacts, six steps in chapters on biological and historic/archeological impacts, and only four in his chapter on socioeconomic impacts. However, the logic of the framework is essentially the same from one chapter to the next.

Canter's ideal impact assessment methods at steps 5-7 of his framework employ state-of-the-art, quantitative prediction models. He is most comfortable when prescribing methods from his specialties within environmental engineering. For example, Canter's (1977: 77-81) methodology for estimating air pollution impacts from a stationary source begins with the collection of meteorological data, particularly mean mixing zone altitudes and the site's "wind rose." (A wind rose depicts frequencies of wind speed from each of the primary compass directions, e.g, north, northeast, etc.) The project's mesoscale impact is calculated as a simple percentage increase in its estimated annual air pollutant emissions over the existing total of emissions within the region. The project's microscale impact is calculated by modeling the ground-level concentrations of air pollutants caused by the project's emissions. The model relies on an exponential function relating forecast concentrations at a given point to the rate of emissions from the project's smokestack, its effective stack height, mean wind speed, atmospheric stability, and distance. The model can use the results from this function to forecast maximum concentrations at any point downwind from the plant. This and Canter's other methods for estimating air, water, and noise pollution impacts are fairly well understood and rely on standard formulae and measures.[5]

Canter is on less solid ground when dealing with biological and especially socioeconomic impacts. He continues to advocate sophisticated analysis: "Quantify [biological] impacts where possible, and qualitatively discuss the implications of the remainder" (p. 146); "[socioeconomic] changes should be quantified where possible and qualitatively described as a minimum" (p. 164). However, he cannot teach his readers handy algorithims for quantifying biological and socioeconomic impacts, as he had been able to do with air, water, and noise pollution, because he is out of his field. (Canter's chapter on archeological and historical inventories technically falls within the social category, but constitutes a very narrow exception to this point.)

A good prescriptive work on quantitative biological impact assessment is a most rare species in this literature, but socioeconomic methods do exist. Economist Larry Leistritz and sociologist Steven Murdock (1981), for example, are schooled in these methods and able to prescribe the theories, variables, and assessment techniques applicable to economic, fiscal, public service, demographic, and other

12

social impacts--all within an assessment approach broadly comparable to Canter's. (Also see DeSouza 1979 and Freudenburg 1986.) Economists, needless to say, can write algorithims every bit as complex as, or more complex than, air and water pollution models.

Canter's and others' frameworks in the prescriptive environmental assessment literature naturally borrow several steps from the standard format of EISs as specified by NEPA section 102(2)(C) and CEQ guidelines (and described in detail in Chapter 2). Canter pays particular attention to the description of the preproject environment; for example, two steps in his framework parallel that standard chapter in EISs. However, the state-of-the-art prescriptive literature goes well beyond the simple EIS format in drawing a model for environmental analysts. It adopts NEPA's rationalist format of analyzing the matrix of consequences of a proposal and its alternatives. More important, methodologists like Canter, Leistritz, and Murdock equate good environmental assessment with the use of quantitative prediction models. That is, they effectively subscribe to the paradigm of Ed Quade and the Rand systems analysts, which is that the use of empirically validated quantitative models is the key to good decisionmaking.

CONTENDING NEPA REFORM MODELS

The National Environmental Policy Act's impact on federal agencies has been more compelling than the creation of one more class of decision document in a federal paper-work system that in 1970 already contained a plethora of such documents. The writing of EISs is embedded in a process that places significant demands on both internal agency resources and agencies' external relations. EIS procedures have changed slightly since 1970, but NEPA review may entail seven distinct steps: preparation of a short environmental assessment, completion of a so-called scoping process, defense of the proposal at public participation meetings, analysis leading to a draft EIS, interagency and public review of the DEIS,[6] revisions in the proposal and filing of a final EIS, and final review before the filing of an official decision. These steps, of course, occur in addition to any standard intraagency decisionmaking steps or post-FEIS administrative appeals or litigation.

The multifaceted nature of the NEPA process has led to several views of how the process could improve or has improved federal natural resources decisionmaking. These

13

views are not wholly incompatible; in fact, resource agencies' implementation of the NEPA process has involved activities compatible with each view. Nonetheless, the process incorporates decisional requirements that are grounded in the competing paradigms of public decisionmaking discussed earlier. Thus, federal agency officials have often felt awkward attempting to execute a package of reforms whose components are based on such divergent assumptions about the nature of public policymaking.

Rational Models

As a product of the rationalist reform movement of the 1960s, rationalist decision theory provides an initial, logical view of EISs as tools of decision reform. The core chapters of these documents incorporate the middle steps in the classic rational decision sequence—identification of alternatives and analysis of their consequences. Since EISs are the most visible product of the NEPA process and the act's "action-forcing mechanism," these documents naturally invite evaluation under the standards of the rational model.

There are three distinct versions of the rational model. The most rigorous is represented by the position, forcefully argued by Keith Caldwell (1982: 2), that NEPA required agencies to focus "scientific knowledge and method" on resource policy choices. In other words, agency administrators' decisions must be guided by state-of-the-art scientific understandings of environmental systems. (Also see Bartlett 1986.) This rational-scientific model requires an official to construct a holistic theoretical framework, based on the best work in many relevant disciplines, of the changes that could be induced in some system by an agency action. This is a demanding, conservation-oriented model. Scientist-conservationists from George Perkins March to Rachel Carson and Barry Commoner have warned of the perils posed by resource development and argued that a wise, science-driven decisionmaker would eschew any environmentally destructive project.[7]

Only slightly less rigorous is the model of the systems analysts and assessment engineers like Canter. Their approach is that of a calculating technician who systematically forecasts N impacts and then aggregates these forecasts to reach a bottom-line determination of each alternative's net benefits. What this rational-comprehensive approach may sacrifice in sophisticated scientific theory—and the sacrifice should be minimal—it makes up for with thorough accounting.

The legal requirements imposed on EIS writers are not quite as strict as those of the scientific and systems analysis versions of the rational model. The legal standards laid down in precedential court decisions interpreting NEPA are not precise. The courts demand objective "good faith" analysis and "full disclosure" of all impacts reasonably anticipatable by agency officials. However, they have declined to require perfect analysis; in legal jargon, impact analysis is limited by a "rule of reason" in which the technical complexity of EIS analysis must be appropriate to the purposes of the document. Judges expect the documents simply to inform so-called reviewing decisionmakers of expected consequences. This legal standard, like many other areas of judge-made law, is a common-man standard. Ordinary people reading an EIS may not demand sophisticated calculations, but they will be most aggravated if they believe EIS writers are dissembling. Thus, the minimum legal version is a rational-objective model holds that EISs should be technically informed, reasonably thorough, and above all else, unbiased.

The Internal Reform Model

NEPA also called on federal agencies to use an "interdisciplinary approach" in decisionmaking; even without such an explicit directive, agencies found that interdisciplinary teams were a practical necessity in EIS writing. The interdisciplinary requirements of the NEPA process lie at the root of the second major model of NEPA reforms. Among the charges of the act's proponents against federal resource managers was that any given agency's staff was usually dominated by people from a single core profession who saw the world through their own discipline's eyes to the exclusion of other (i.e., ecological) concerns, and whose professional blinders tended to exacerbate the most damaging features of an agency's resource development mission. For example, the civil engineering training of officials in the Corps of Engineers supposedly led them to focus their attention on constructing water resources structures and to ignore project consequences and externalities that were not the business of civil engineers. NEPA, so the "internal reform" argument goes, brought new cadres of officials into federal agencies to implement the act's mandate and staff EIS-writing teams (Wichelman 1976: 279-283; Culhane 1974: 35-36). These biologists, landscape architects, archeolo-

15

gists, and others who were not from an agency's traditional core profession became, in effect, in-house advocates for environmental consciousness during agency deliberations.

In principle, a broadening of the disciplinary base of federal agencies is logically consistent with the most rationalist vision of NEPA. Indeed, interdisciplinarity is an essential element of the holistic ideal of the rational-scientific reformers. However, both the stances of the proponents of the internal reform model and its theoretcal roots make it a clear alternative to the rationalist view. Sally Fairfax, who coined the term "external reform," regards NEPA's mandate to write comprehensive EISs as a "disaster" (Fairfax 1978, Fairfax and Andrews 1979). Serge Taylor (1984) also subscribes to the internal reform position, criticizes the rational-scientific view of the NEPA process, and recognizes that NEPA involves fundamental changes in intraagency politics and dynamics.[8]

The internal reform position represents a special instance of the approach to decisionmaking, begun by Herbert Simon (1947), that Allison (1970) calls the "organizational process model." This model essentially holds that administrators' decisionmaking is most efficient when officials follow standard operating procedures derived from satisficing, not rational-optimizing, choices. A key to understanding these officials' motivations is that, whether through prior education (e.g., in the discipline of their professional work) or on-the-job socialization, officials develop strong identifications with their agency's way of doing business. The internal reform scenario thus predicts that, by importing people from diverse professions, an agency can expand its standard procedures to include the routines of the new professionals. Those new people will be just as strongly motivated by their professionalism as the people from the agency's traditionally dominant profession.

The External Reform Model

The third reformist vision of NEPA focuses on external pressures reinforcing environmental values in agency decisions. Before NEPA, agency decision processes tended to exclude actors other than agency officers and project beneficiaries (Reich 1962). However, CEQ's NEPA guidelines, by forcing agencies to circulate draft proposals for public comment, allowed any person to become an actor in routine federal resources decisionmaking. That person could be a private citizen like Thomas Horobnik, a Missoula, Montana,

16

retiree who commented on virtually every Forest Service EIS released during the mid-1970s. It could be, and commonly was, a citizens group that objected to the proposal. Or it could be another federal or state agency whose mission was affected by a lead agency's project but which would not have been afforded an opportunity to effectively express its reservations before NEPA. NEPA, in other words, created a historically unique mechanism for interest group, inter-agency, and intergovernmental pressure on federal agencies (Andrews 1976, Liroff 1976, Friesema and Culhane 1976).

Moreover, soon after the act was passed, it became clear that the federal courts would be receptive to NEPA liti-gation. The combination of environmentalist litigation and judicial interpretation made NEPA a much more potent force for change in federal policymaking than the act's sponsors had ever anticipated (Liroff 1976). By the mid-1970s, federal resource administrators suffered from a common nightmare that some environmentalist would find a judge, obtain an injunction, and bring their programs to a crashing halt. NEPA suits actually proved to be less effective than suits under other statutes in forcing cancellation of projects, as opposed to simply delaying them interminably (Wenner 1982). Nonetheless, NEPA suits were a major source of external pressure on agencies, as well as a threat that made agencies respect routine public and interagency comments on EISs.

The external reform model is obviously grounded in the political philosophy associated with Lindblom's (1965) "partisan mutual adjustment" and Allison's (1970) "governmental politics" model.[9] The external reform and rationalist models are thus at cross-purposes. Actors in politicized decision settings find comprehensive analyses difficult because the first step in the rationalist decision sequence, acheiving goal consensus, is so elusive. The rational-comprehensive decision model, as Charles Anderson (1979: 712) put it, is like the recipe for rabbit stew that begins, "First, catch the rabbit."

While proponents of the internal-reform and external-reform models believe rationalist EIS analysis is rarely possible, some of them disagree over the compatibility of the internal and external visions of NEPA reform. In a controversial 1978 Science article, Sally Fairfax argued that a debate existed between the proponents of the internal-reform model, of whom she was one, and the advo-cates of externally pressured reform. The latter course of action, she (1978, 1979) argued, involved a disastrous waste of resources in conflicts over meaningless documents when it was imperative to improve the intraagency deliberations that actually determined policy. However, most other proponents of both the internal-reform model (Wichelman 1976, Taylor 1984) and the external-reform model (Culhane 1974, Liroff 1976) see the two types of changes as complementary and mutually reinforcing. That is, a steady external threat enhances the clout of environmental staffs inside the agencies, and the goal of outside critics of agency programs is not to sue and criticize, but to induce agencies to respect environmental consequences.

PLAN OF THE BOOK

The specific research questions addressed in this book are directed at the dispute between rational and arational decision theorists. An examination of decisionmaking must confront the rival models on the ground of their opposition--the cognitive psychology and decision processes. Do decisionmakers use models to comprehensively estimate consequences of alternatives and then choose the optimum, or do they decide through satisficing choices within a framework of agency standard procedures and bureaucratic momentum, or do they decide as a result of negotiations and political pressures? Only one study, to our knowledge, accepts this challenge. Nienaber's and Wildavsky's (1973) study of the implementation of PPB in public lands agencies demonstrated that PPB fell far short of its rationalist promise and failed to supplant traditional arational budgetary routines. About the time Nienaber's and Wildavsky's book was published, the government came to the same conclusion and informed federal agencies that they no longer need spend time preparing PPB paperwork.

Our study of NEPA evaluates rationalism on its own ground. This study cannot compellingly test both rival paradigms of decisionmaking. The research questions described below are directed almost exclusively at the rationalist vision of the NEPA process. However, this study is not just a "strawman" exercise. Some readers may feel that testing the technical rationalism of federal decision-making is about as fair as challenging one's grandmother to an arm-wrestling contest. We are not proponents of the rationalist model, but rationalism's advocates argue that rationalist policy prescriptions are both eminently desirable and implementable in practice. In the 1960s these people advocated PPB. In the 1980s they advocate cost-benefit analysis of federal regulations. As detailed above, most of the prescriptive literature on environmental assessment subscribes to the rationalist ideal. NEPA, moreover, mandated the adoption of rationalist procedures at local and regional levels of administrative decisionmaking, where the prognosis for a rationalist reform should theoretically be better than at the macropolitical level, where PPB budgeting foundered.

Stated simply, this book examines the analytical quality of the environmental impact statements that lie at the heart of the NEPA process. A good EIS prediction should be technically sophisticated and correct. These criteria of sophistication and correctness translate into two primary research questions, plus an interesting intermediate topic. The bottom-line question is, how accurate are the predictions of environmental and socioeconomic impacts that one finds in the average EIS? The primary objective of this study is to determine the match between a set of forecasts in a representative cross section of federal EISs and the actual impacts that occurred following project implementation. As a preliminary step, one must systematically inventory and analyze the forecasts in the sample of EISs. This content analysis provides evidence about the technical sophistication of EIS predictions. In addition, before evaluating whether an EIS's forecasts are correct, one must determine when the EIS's proposed action was implemented to know whether sufficient time has passed for predicted impacts to materialize and whether the project implemented was the same as the one proposed, since predicted impacts are logically a function of the characteristics of the proposed action.

Chapter 2 provides additional background information on NEPA and its implementation. Our interpretation of this historical material is based on our familiarity with the NEPA process during its first decade and a half and with the literature on NEPA, rather than on any primary data.

Chapter 3 describes the study's sample. First, a set of EISs filed between 1974 and 1978 was randomly drawn from the U.S. Environmental Protection Agency's (1980) list of EISs. Second, we determined if and when each EIS's proposal had been implemented. This determination provided a case study of the implementation status of a random sample of federal proposals. Third, a field sample of 29 projects was selected. The sample EISs were written by 13 lead agencies, including most major EIS-writing agencies. The sample contains similar numbers of cases within five major classes of projects—water resources, public lands, energy, highways, and urban or building projects.

Chapter 4 presents a content analysis of the forecasts within the field-sample EISs. A "forecast" is defined as any passage in a final EIS about the consequences of the proposed action. The research team used a pretested protocol to encode 12 characteristics per forecast. This chapter's conclusions about the sophistication of EIS forecasts--or the lack thereof--bears directly on the question of the technical rationality of U.S. environmental assessment.

Chapters 5, 6, and 7 confront the principal concern of the study, the accuracy of EIS forecasts. The study's basic model of impacts is a pre- and postproject interrupted time series design, in which the project serves as the inter-ruption. (See Cook and Campbell 1979.) This model of impact accuracy requires an ideal forecast to provide preproject observations on an affected parameter for a baseline trend and postproject forecast values for an impact trend. Impact accuracy is then determined by comparing actual postproject observations with the forecast and preproject trends. We originally proposed to analyze the data using a pooled cross-sectional time-series statistical model. However, that technique proved inadequate in the face of the complexities and limitations of EIS forecasts, discussed in Chapter 4, and the varied nature of available field data.

Chapter 5 reviews the our fieldwork experience in searching for data on project impacts. Field teams arrived at each project site with a list of candidate fore-casts in seven categories of forecasts: physiographic, biological, economic, and social impacts, plus mitigation measures, project objectives, and a controversial impact. The teams attempted to acquire time-series data bearing on predicted impacts from existing records. The teams also conducted 120 interviews with informants familiar with the projects to identify significant impacts that EIS writers failed to anticipate and explanations about patterns in the quantified data. Good quantified data often proved to be unavailable on a given impact, and those data that were available tended to be unevenly distributed among projects and impact categories. This experience represents a substantive finding in its own right. That is, in contrast to the self-evaluating decisionmaker of the systems analysis model, implementing officials in federal agencies do not possess adequate data on the consequences of their projects.

Chapter 6 provides examples of the auditing scheme used to estimate forecast accuracy. In addition to the quanti-fied data that are available, nonquantitative evidence is

20

quite adequate for evaluating certain forecasts. Such a varied range of information, of course, cannot be evaluated in a pooled statistical model. Instead, the analysis employs a case-by-case classification of each forecast impact. Fifty-eight example cases and their classifications are used to explain this complex classification scheme.

Chapter 7 summarizes our audit of the predictive accuracy of the 239 forecast impacts in the study. Essentially, this chapter tells a "good news and bad news" story. The good news is that, when we can draw firm conclusions about the match between forecasts and impacts, EIS forecasts are correct much more often than they are incorrect. The bad news is that intellectually satisfying conclusions often cannot be drawn for a variety of reasons, most of which reflect poorly on the comprehensiveness of the sample EISs.

Chapter 8 concludes the study by returning to the contending rational and arational visions of NEPA reforms and decisionmaking in general. The study's data, for the most part, can only test the rational vision and, as we have hinted above, find it wanting. Nonetheless, we offer some observations on the strengths and weaknesses of the NEPA process and on alternative visions of the NEPA process. Our stance on this subject will not surprise anyone who is familiar with our prior research into the NEPA process, but with this study we believe we have a much firmer empirical base for that stance.

NOTES

1. The number 10,475 equals the total of final EISs plus draft EISs for which no final EIS had yet been filed, on file as of May 1980 (EPA 1980). For a variety of reasons some EISs that agencies write are never officially filed with the EPA or CEQ. The 1981 final EIS on Nevada-Utah MX basing, for example, was printed but never filed because President Reagan abandoned that plan.

2. For more modern examples, see Cohen, March, and Olson (1972), Padgett (1980), Lustick (1980), and Olson (1984).

3. NEPA's statutory provisions and administrative implementation will be covered in more detail in Chapter 2. Among the best books on NEPA, Liroff (1976) and Andrews

(1976) each devote a chapter to the passage of NEPA, and Andrews (1976), Liroff (1976), and Taylor (1984) examine NEPA implementation by major federal agencies.

4. See, for example, the handbooks by Baker, Kaming, and Morrison (1976), Burtchell and Listokin (1975), and Orloff (1978). The assessment literature also contains a number of unusual works, such as Medford's (1973) book that touts environmental assessment to a European audience. Some of the prescriptive handbooks, such as that of Cheremisinoff and Morresi (1977), fall noticeably short of the quality of the Canter and Leistritz and Murdock works discussed in the text.

5. Canter's air quality methodology assumes that the proposed project is a standard air pollution stationary source, but only a minority of all EISs are written on traditional stationary sources. He does not cover the air pollution impacts of the large group of highways and their "mobile source" vehicles.

6. Following the language of section 102(2)(C), we generally use the term "interagency review" to refer to comments by federal, state, and local government agencies or officials. In our experience (Friesema and Culhane 1976), there are noteworthy differences in the roles and influence of public, federal agency, and state or local commenters.

7. On the similar roles of scientists in the nine-teenth-century conservation and contemporary environmental movements, see Culhane (1981: 3-10).

8. Taylor (1984) generally uses the phrase "science model" of EIS analysis to encompass all three versions of the rational model, plus interdisciplinarity. His use of the term is unique, and distinct from the linkage of science and the NEPA process advocated by Caldwell (1982).

9. The dynamics of external review of EISs can be distinguished from Allison's governmental politics model because the latter is specified in ways appropriate for use in foreign policy analysis (Culhane 1987).

2
The Evolution of the NEPA Process

The National Environmental Policy Act (NEPA) has had a major effect on the normal routines of government bureaucracies, however it may have affected the human environment. The bureaucratic impacts attributable to NEPA include impacts on agencies' scientific missions, planning processes, decisionmaking routines, and ties to citizens and interest groups. The many consequences for agency operations that are attributable to NEPA could not have been predicted, or at least were not predicted. NEPA's effects astonished just about everybody. The bureaucratic and scientific consequences of NEPA emerged over the early years of implementation of the act. During this period, agencies were frequently shocked into organizational action by a court decision interpreting some aspect of the NEPA requirement, and they have sometimes been similarly shocked by citizen or intergovernmental responses to EISs they have prepared. But these externally induced shocks led to the development of common agency interpretations, which in turn led to standard operating procedures for implementing the act. This standardization has been fostered by Council on Environmental Quality guidelines, and somewhat later by CEQ's compulsory regulations.

Many standard agency procedures and interpretations, as they have emerged, have had substantive consequences. They have defined the types of analyses that needed to be conducted and have structured what has actually gone into EISs. Many of these substantive expansions and alterations in the meaning of NEPA affect this study. Moreover, the time boundaries within which our sample of EISs were selected are based on our judgment about bureaucratic standardization in implementing the act. We believe that standardization of

23

implementation was basically complete by 1974, four years after passage of the act. We also believe that, despite what is generally perceived to be a significant revision of NEPA implementation requirements fostered by new CEQ regulations in 1978, those new regulations were basically codifying standard procedures, particularly concerning what needed to be included in an environmental impact statement and how it was to be included. So the procedures and interpretations that had become standard across agencies by 1974 still, in large measure, structure the preparation and use of environmental impact statements. Those standardizations that apply to the study are discussed in the balance of this chapter. The standard agency procedures are all, of course, within parameters established by NEPA statute and then within the narrower parameters imposed by the Council on Environmental Quality and the political process structured by the CEQ. As understanding of the statute itself and the political process it created are essential to an understanding of the subsequent bureaucratic standardization.

THE STATUTE

The National Environmental Policy Act contains three major sections. Section 101, together with the rest of statutory purposes in Section 2, constitutes a "Declaration of National Environmental Policy." This declaration, as is common in such statutory passages, sets forth a long list of broad, ambitious goals:

. . . [to] encourage productive and enjoyable harmony between man and his environment: to promote efforts which will prevent . . . damage to the environment . . .; to enrich the understanding of the ecological systems . . [§2];

The Congress, recognizing the profound impact of man's activity on the interrelations of all components of the natural environment . . . declares that it is the continuing policy of the Federal Government . . . to use all practicable means . . . to foster and promote the general welfare, to create and maintain conditions under which man and nature can exist in productive harmony, and fulfill the social, economic, and other requirements of present and future generations of Americans [§101(a)];

24

. . . it is the continuing responsibility of the
Federal Government to . . . fulfill the responsibilities of each generation as trustee of the
environment for succeeding generations, . . . [to]
attain the widest range of beneficial uses of the
environment without degradation, . . . or other
undesirable and unintended consequences . . . and . . .
[to] enhance the quality of renewable resources
and approach the maximum attainable recycling of
depletable resources [§101(b)].

NEPA's sponsors recognized that lofty statutory
declarations of policy would have little effect on agency
decisionmakers. Following pivotal committee hearing
testimony by Professor Lynton Caldwell of Indiana University, Senator Henry Jackson (D., Wash.) and his key interior
committee staffers drafted section 102 to force federal
agencies to observe the policies of section 101. This
action-forcing mechanism was a requirement that agencies
write a "statement of environmental findings." Senator
Jackson and his staff were influenced by President Johnson's
order two years earlier that all agencies prepare PPB
documents for inclusion with their annual budget requests
(Liroff 1976: 16-18). They believed that NEPA's environ-
mental-findings requirement presented an opportunity for
Congress to require that it be given choices among ranges of
alternatives, just as PPB provided the executive branch and
especially the Office of Management and Budget (OMB) with
such choices.

The environmental impact statement requirement of
section 102(2)(C), as finally passed, reads as follows:

. . . (2) all agencies of the Federal Government shall
. . . (C) include in every recommendation or report on
proposals for legislation and other major Federal
actions significantly affecting the quality of the
human environment, a detailed statement by the
responsible official on--
  (i) the environmental impact of the proposed
  action,
  (ii) any adverse environmental effects which cannot
  be avoided should the proposal be implemented,
  (iii) alternatives to the proposed action,
  (iv) the relationship between local short-term uses
  of man's environment and the maintenance and
  enhancement of long-term productivity, and
  (v) any irreversible and irretrievable commitments

of resources which would be involved in the proposed action should it be implemented. Prior to making any detailed statement, the responsible Federal official shall consult with and obtain the comments of any Federal agency which has jurisdiction by law or special expertise with respect to any environmental impact involved. Copies of such statement and the comments and views of the appropriate Federal, State, and local agencies, which are authorized to develop and enforce environmental standards, shall be made available to the President, the Council on Environmental Quality, and to the public as provided by section 552 of title 5, United States Code [the Freedom of Information Act], and shall accompany the proposal through the existing agency review processes.

Jackson and the section's other drafters envisioned NEPA's environmental statements as relatively modest reports that would accompany a proposal and other documents on the proposal through its review by OMB examiners, and assumed all this material would ultimately reach the congressional authorization committees. Because of court cases, administrative responses to the cases, and other pressures, EISs have instead become the centerpieces of a multistage NEPA process. The EIS documents themselves have grown so that they characteristically run into the hundreds of pages.

NEPA's primary sponsor in the House, Representative John Dingell (D., Mich.), added the third major element of the act. Title II, that is, sections 201-207, created the Council on Environmental Quality as a staff agency within the Executive Office of the President. The CEQ had a variety of historical precedents. Its most recent predecessor was the Council of Economic Advisors, the small group of professional economists that has generally played a major role in formulating presidents' macroeconomic policies. The idea of a central conservation commission, however, dates at least from the National Resources Planning Board, a New Deal management innovation that foundered during World War II (cf. Clawson 1981). The management principle underlying an agency like CEQ is that control of important bureaucracy-wide functions must be vested within the executive staff of the president, and NEPA implied that environmental reform was indeed such an important function. On paper, however, CEQ's functions were quite vague--preparing an annual report on environmental quality and advising the president. Even though the act did not do so, in March 1970 President Nixon gave CEQ its natural responsibility for overseeing the NEPA

26

process, including, of course, the production of EISs by federal agencies (E.O. 11514).

In addition to the three primary elements of the act, NEPA contains three minor sections, which follow section 102 within title I. Section 103 required federal agencies to report by July 1971 on any inconsistencies between their existing statutory missions and NEPA's purposes. Few agencies complied with this requirement effectively or on time, and little attention was paid to the nonimplementation of the requirement. Section 104 is a technical provision that confirmed NEPA was not intended to hamper compliance with prior pollution control or environmental quality standards. Section 105 stated, "The policies and goals set forth in this Act are supplemental to those set forth in existing authorizations of Federal agencies." This section was inserted to placate congressmen who, while they did not fully comprehend the act's implications any better than its sponsors, nonetheless wished to prevent NEPA from interfering with the resource development programs they supported.

THE NEPA PROCESS

This relatively short statute spawned numerous court cases, as well as a variety of organizational responses to its vague mandates. The statute itself and the messages coming out of the court cases were hard to interpret. In a series of 1971-1973 decisions, federal courts held agencies to strict compliance with NEPA's procedural mandate.[1] All the precedential NEPA decisions of the early 1970s were handed down by lower federal courts, with the D.C. Circuit playing a particularly important role, notably in its Calvert Cliffs (1971) decision. By the 1980s, the U.S. Supreme Court had set a pattern of interpreting NEPA narrowly. In eight major decisions from 1976 to 1983, the Supreme Court read NEPA's mandate restrictively and reversed lower court—especially D.C. Circuit—decisions holding that federal agencies had insufficiently complied with NEPA's mandate.[2] However, by the end of the 1970s the NEPA process had become a set of standard operating procedures within the federal bureaucracy and had been legally entrenched in formal regulations, so those restrictive rulings did not substantially affect the NEPA process. Instead the NEPA process was developed, in the main, to solve the administrative and political problems faced by the agencies in adjusting to very complicated pressures.

The standardization of NEPA was fostered by the Council on Environmental Quality, which gave agencies a good deal of informal advice. CEQ issued guidelines about preparing EISs (CEQ 1970, 1971, 1973) that followed and adhered to the NEPA decisions and evolving agency practices. The court cases, as reflected in and expanded on by CEQ regulation, created a political process affecting all federal agencies. The CEQ contribution to the evolution of the EIS requirement was most pronounced in the introduction and evolution of public participation elements into agency compliance with the act.

CEQ's initial guidelines interpreted the clause in NEPA that "copies of such statement and the comments . . . of the appropriate Federal, State, and local agencies . . . shall be made available to the President, the Council on Environmental Quality, and the public" to obligate lead agencies to circulate a draft EIS for comments before writing a final EIS. In so doing, CEQ (1970) added an element to the NEPA process that had not been specified by the act's congressional sponsors. It created an opportunity for review of proposals by "the public," as well as by the federal agencies and state and local governments specified in section 102(2)(C). That "public" clearly included environmental and citizens groups that often opposed resource development programs and had seen themselves as excluded from insider decisionmaking about those programs. By CEQ guidelines, later made into requirements, a lead agency must allow a minimum of 45 days for public, inter-governmental, and interagency review of a draft EIS. Following this comment period, the agency must evaluate these comments, respond to them, revise the DEIS analysis as necessary, and write the final EIS.

Rather than channeling public pressures, as the agencies hoped, these participation requirements provided additional avenues for the articulation of public criticism. The act itself and the CEQ oversight of the environmental impact statement process created the boundaries within which agencies could work out their procedures. Those procedures were basically in place within a few years of the passage of the act. The procedures generally required some clear stages, beginning with the decision over whether an EIS needed to be prepared in the first place. If a decision was made that an EIS was needed, then an interdisciplinary team was assigned responsibility for preparing the analysis. Once an agency was satisfied with the contents of that EIS, it was released as a draft for public and interagency review, as indicated by CEQ. After the comments were in, the agency prepared a final EIS that responded to the comments it received on the draft. The distribution of the final EIS sometimes ended the "NEPA process," but often it did not. The EIS often became the basis for a continuing controversy over a project in some other forum.

There have been some changes in NEPA procedures and terminology since the early 1970s, but they have not fundamentally altered the internal agency interpretations or processes for producing environmental impact statements.

Standardization of the Definition of Need for an EIS

NEPA requires agencies to prepare environmental impact statements whenever a major federal action may have serious consequences for the human environment. But how can an agency decide whether an action may have a major impact on the human environment except by conducting an environmental analysis? Agencies have broken this circularity by defining certain types of actions as automatically requiring an EIS, certain other actions as automatically not requiring an EIS, and an intermediate class of activities that require some type of of preliminary assessment to determine whether a full-blown environmental impact statement is required.

In its 1978 regulations, CEQ standardized this system by imposing terminology and procedures on all agencies. Before 1978 individual agencies called their preliminary environmental documents "environmental analyses," "environmental reviews," "statements of environmental concerns," and other names. Since 1978 these intermediate analyses have been called "environmental assessments." An agency must now follow such an analysis up with either a decision to prepare an EIS (which it rarely does) or a "Finding of No Significant Impact," or FONSI. Before 1978 this finding had been called by various names, frequently a "negative declaration."

It is important to note that the CEQ regulations of 1978 codified and standardized what the agencies were generally doing anyway. And a decision to prepare an EIS has rarely been based on an environmental assessment (EA). Rather, agencies have adopted rules of thumb and incorporated these practices into their own regulations implementing NEPA. These rules of thumb are usually based on the magnitude of a project, rather than on any measurement of its possible environmental impact. Thus, if some HUD loan guarantee involves 500 housing units or a highway involves a certain number of miles of new construction, an EIS will be prepared. Because agency missions vary, there are still some differences in agency determination of the need for an EIS. A much smaller federal action may trigger an EIS if it is likely to affect a particular environmental value (a small road through a national park or a small housing project to be located in a valuable wetland).

Because much of the early NEPA litigation concerned when an EIS was required, agencies set their standards and processes fairly soon after the passage of NEPA. Since that time agency thresholds have risen a bit, so a somewhat larger project or a larger possible impact may be needed to trigger a full-blown EIS in 1987 than in 1973. There is still a good deal of elasticity in the process, however. A highly controversial project is more likely to stimulate an EIS than a less controversial one.

Perhaps as many as three of the EISs in our sample would not have been required if the projects had been proposed in 1987. While most agencies are writing fewer EISs in 1987 than they did during the period covered by our research, this decline in EIS production is attributable to other factors besides a raising of thresholds. Agencies have eliminated their backlogs of authorized projects. There are far fewer new starts on many types of projects (water resource developments, interstate highway segments, etc.). However, one agency is producing far more EISs than it was during our sample period. As a result of legal arguments, the Bureau of Land Management is now preparing EISs to accompany its basic land use planning documents, called Resource Management Plans. During the time of our sample, the BLM maintained that the RMP (then called a management framework plan) was a conceptual document and that NEPA compliance should come as part of the action elements of that plan, not as an accompaniment to the plan. Since the late 1970s the agency has had to alter its interpretation, and it is now interpreting the NEPA requirement in much the same way as the other federal land managing agencies.

Expansion of the Meaning
of the Human Environment

The National Environmental Policy Act, as interpreted, requires a federal agency to prepare an environmental impact statement on any project that might have a major impact on the human environment. But what is the human environment? It seems clear that the legislative creators of NEPA were primarily concerned with government project impacts on biotic or natural systems, but also (more vaguely) with the generation of pollutants and other distasteful or dangerous byproducts and with any disruption of living systems that might affect human health and satisfactions. They probably did not envision NEPA as being heavily oriented toward the social and economic impacts of projects.

Nonetheless, external challengers to proposed government projects have frequently jumped on the vague term "human environment" to propose that projects not proceed until issues of social and/or economic impact have dealt with in an environmental impact statement. These proposals have met a mixed reception by the agencies and the courts, or so a study of the case law would suggest. The courts and CEQ (1978) have generally tried to limit the meaning of "human environment," and thus the requirement to prepare an EIS, to impacts on natural systems or to the direct human consequences of alterations in natural systems.

Despite understandable agency interest in restricting the scope of NEPA, reinforced by CEQ guidance and a string of court decisions, agencies have by and large considered social and economic impacts a routine part of EISs. Indeed, the social or economic impacts of projects have often been the most controversial prospective impacts, and have often received the most attention in the environmental assessments that have been conducted by federal agencies. This was already a clear pattern within a couple of years after the passage of NEPA.

Social impact analysis rapidly evolved as a part of the NEPA compliance procedures of federal agencies. The earliest EISs considered the possible impacts of projects on historical properties and archeological materials, which might be considered social analysis. Such impacts were incorporated into EISs because the agencies were already required to consider them under other legal mandates. The social impact analyses quickly expanded beyond this base. As large resource development projects were proposed, particularly in remote and undeveloped areas (Alaska and the intermountain West), it became necessary to consider their impact on particular human populations, such as American Indians or Eskimos, or on rural and smalltown life. It was practically (if not legally) necessary to consider these impacts because everyone believed they were very important. Whatever an agency might have wished to do, the public and inter-agency review process raised the social impact issues. When EIS respondents, including key constituent groups, demanded that agencies be concerned about social impacts, the agencies complied. NEPA, as interpreted, also required agencies to respond to public and interagency reviews.

Soon after this pattern of considering a broader array of social impacts began, it spread across agencies and achieved the status of standard operating procedure. This probably occurred for a variety of reasons, including the fact that bureaucracies frequently handle uncertainties by following precedents. Associations of social impact assessors soon appeared, with all the appearance of nascent professions--journals, conferences, "networks," and attempts to create standards that would limit entry to the profession. Social scientists became routine members of interdisciplinary EIS teams. Thus, the incorporation of social impacts into EISs is largely a product of agency practice, not legal interpretations of NEPA requirements. The operating rule is that prospective social impacts, even if important, will not in themselves trigger an EIS, but if an EIS is required, prospective social impacts are just as important as biological or physiographic ones.

The routine consideration of economic impacts of projects in EISs followed a similar evolution. While the courts indicated from time to time that agencies need not incorporate economic cost-benefit analyses in their EISs, agencies quickly began to consider economic impacts as routine parts of their environmental assessments. However, investment economics and sponsor's profits are rarely mentioned in EISs, even though they naturally play a major role in many project proposals. Some special considerations probably fostered the introduction of analyses of economic impacts. Most important, economic development or betterment was often the reason a project was proposed in the first place, so it was natural, in trying to justify a proposed federal action, to include the (usually available) economic projections that had been made. This tendency to incorporate economic analyses into the project justification was reinforced as handbooks and other guides to EIS preparation began to appear and suggest that the best EIS predictions and analyses were quantitative. It was easier to respond to inducements to present quantitative data on economic projections than it was to do so on most other possible impacts.

The rapid routinization of the consideration of social and economic impacts was clearly aided by agency perceptions of great organizational uncertainty engendered by early court cases interpreting NEPA. Since agencies feared that courts might stop their operations for failure to adequately comply with a vague and expanding EIS requirement, it often seemed prudent to adopt an open definition of the "human environment." Those short-term tactical judgments contributed to the emergence of standard operating procedures in which social and economic impacts could be normal concerns in preparing EISs.

Evolution of Responsibility for Preparing EISs

In the early attempts to comply with NEPA, agencies prepared EISs, or had them prepared, in a wide variety of ways. When the EIS requirement was treated as basically a statement for the files that an agency had considered the environmental consequences of its actions (what the Corps of Engineers labeled a "five point statement" after the five identified points in the law), some agencies had the responsible operating official, whatever that person's scientific or professional credentials, prepare the "paperwork." Some agencies asked applicants to submit an environmental report as part of an application. But as legal action interpreting NEPA began to surface, a consistent agency movement developed to transfer responsibility for EIS preparation from operating officers and offices to some specialized offices or teams.

The creation of specialized EIS teams or offices occurred most readily in agencies that already had fairly extensive planning and research activities. Thus, specialized multidisciplinary teams or environmental offices and more scientifically complex analyses first appeared in EISs of agencies such as the Corps of Engineers, Forest Service, and Atomic Energy Commission, with large organizational resources to draw on.

During the time of our sample, there was still some variety across agencies as to where EISs were prepared. The Federal Highway Administration was the official lead agency on a great many EISs, but these EISs were prepared by the state highway departments that had the resources to prepare them. The FHWA basically slapped a new cover on the state product. Some agencies required extensive assessments for applicants, which they then adopted as their own. In modified form, these practices continue to this day.

Despite variations in agency preparation of EISs, there was a consistent movement, across agencies, toward professionalizing that preparation. Interdisciplinary teams of EIS preparers became the norm, whether residing in an agency's planning or research offices, in a separate EIS office, in an applicant's office, or in a set of hired consultants. Those teams were composed of professional analysts rather than operating personnel. While members usually lacked Ph.D.s, teams almost always included leaders and members with specialized advanced training in the sciences and the planning profession. This pattern was well in place by 1973, and has changed only a little since that time.

The creation of interdisciplinary teams was clearly mandated by the language of NEPA itself. As the process evolved and the interdisciplinary teams came to include analysts with formal training in diverse academic disciplines and traditions, and not just intra-office groups, this contributed to another major change in government agencies. Many government agencies had professional staffs that were definitely not interdisciplinary before NEPA. After NEPA, no major federal agency could be dominated by civil engineers, foresters, range scientists, or any other professional specialists with anything like the disciplinary uniformity that had existed earlier. The influence of internal agency "environmentalists" as EIS team members began to affect agency decisions early in the NEPA years. This has continued, as some interdisciplinary recruits have continued their careers within these agencies. This restructuring of agency staffs may be one of the most critical consequences of NEPA.

By 1973 EISs were applied scientific documents, as they are now. Agencies had specialized staff members whose careers were tied to the evaluation of the EISs they prepared. They also had clients and interested publics who had clear expectations about what should be in EISs. These factors made it important for agencies to produce credible scientific documents to accompany their proposals.

Integration of EIS Preparation with Project Planning

Many federal agencies that are required to prepare EISs are also required to prepare fairly elaborate plans in conjunction with their missions. Planning efforts are most pronounced and elaborate in water resource agencies, such as the Corps of Engineers, and land managing agencies, such as the Forest Service and National Park Service. It was quickly apparent that some EIS requirements were interrelated with some agencies' planning requirements. In fact, some agencies, such as the Forest Service, began writing EISs on their plans, as distinguished from specific action elements or decisions. Before long, many of the planning agencies were trying to integrate their planning and NEPA compliance to reduce obvious duplication and unnecessary costs.

In addition, these agencies also began to integrate both their planning and NEPA compliance with their public participation programs. (See Culhane and Friesema 1979.) Public participation programs, consisting of public meetings and a variety of more or less sophisticated participation techniques, became routine features of resources planning during the 1970s. These programs, however, were not mandated by NEPA, even through agencies often cited NEPA as statutory authority for their participation. The Forest Service and Corps of Engineers, for example, instituted their participation programs in 1970 (Mazmanian and Nienaber 1979, Culhane 1981), before CEQ's first NEPA guidelines.

Ultimately most agencies with planning mandates have integrated their planning efforts with the EIS requirements in a number of ways. Some agencies such as the National Park Service, and more dramatically the Bureau of Land Management, started writing EISs on their land use plans rather than limiting them to individual agency decisions. The planning agencies also moved to integrate their planning documents with EISs, so that a single document was prepared and circulated. This process has led to important standardization of processes across agencies.

The effort to integrate NEPA compliance with other planning mandates began shortly after passage of the act, but it took longer to integrate these efforts than to accomplish some other organizational procedures discussed in this chapter. Linking of plans and EISs was well underway during the time of our sample, but it was not complete. One Forest Service EIS in our sample, for example, was accompanied by a separate land use plan when the draft documents were circulated for comment. By the time the final EIS and plan were ready to be circulated, they had been incorporated into a single document. Since the EIS had to describe the proposed action (i.e., the plan), it was obviously efficient to combine these documents.

The NEPA Process and the 1978 Regulations

Court interpretations of the requirements of the National Environmental Policy Act provided the backdrop for administrative standardization. While the courts have provided some confusing guidance to the agencies, this has ultimately led to a much more stable set of legal restrictions on agency activities.

The CEQ has responded to the changing pattern of court decisions in an incremental and minimal way ever since NEPA was passed. The one notable event in this evolution was the preparation of new regulations in 1978. Many of these changes have already been discussed. The regulations basically required the agencies to do what they were already doing, and had been doing since the early days of NEPA. However, the 1978 CEQ regulations introduced two other notable patterns to NEPA compliance.

The first of these innovations occurred at the beginning of EIS preparation, and the second occurred after EISs were fully prepared. Once an agency had administratively determined to prepare an EIS, the next step was to conduct a "scoping" process, which involved identifying the major issues to be considered in the EIS and prioritizing potential issues. Agencies were encouraged to treat this process as one of the interagency and public participation stages of NEPA and to conduct one or more public and interagency scoping meetings. Something akin to scoping was fairly common among agencies even before the 1978 regulations. We know of no studies of the impact of the scoping requirement, but we do not believe it has altered the contents of EISs as a whole to any significant degree.

The other innovation of the 1978 regulations was a requirement that agencies prepare a "record of decision" following the completion and distribution of final EISs. After a modest period of time following distribution of final EISs to allow for public or interagency protest, agencies were to write reports identifying the alternatives they had selected and justifying any elements of decisions in which alternatives that were not the environmentally preferred ones were selected. We are not sure whether this has changed agency decisions significantly, but we are fairly confident that it has not affected the contents of the EISs themselves.

This provision for a post-FEIS record of decisions highlights an important fact about the NEPA process. The final EIS, even when it reveals a clear agency preference, should not be treated as a final decision. The FEIS is, in theory, written to inform some ultimate decisionmakers about the consequences of their choice. In many cases the FEIS seems to be a _pro forma_ exercise because the proposed action is processed mechanically by agency officials through their standard decision procedures without major opposition or controversy. In other cases, however, the commenting pressure continues unabated past the filing of the FEIS and record of decision. Legislators or other authorities are drawn into the controversy, and the proposal is significantly modified, delayed, or even canceled (Culhane and Friesema 1977). That pattern was true before CEQ imposed the record of decision requirement, and it is true today.

In summary, there is a fairly coherent pattern to the production of environmental impact statements. First, the pattern begins with the decision over whether an EIS is to be prepared. This is essentially made on categorical grounds, as reflected in CEQ regulations and agency rules implementing those regulations. Second, after a decision to prepare an EIS is made, the lead agency convenes scoping meetings or otherwise tries to structure project planning by deciding and recording what must go into the EIS and what might be eliminated. Third, using this guidance, an interdisciplinary team divides the analytic and writing responsibilities and prepares a draft EIS. Fourth, the DEIS is circulated for public, intergovernmental, and interagency comments. Fifth, following review of the comments, the team prepares a final EIS. When agency leaders approve the release and filing of the final EIS, a new round of public participation may occur. Sixth, after the agency prepares a record of decision, the formal EIS process is complete. The term record of decision is usually a misnomer because other decisional hurdles usually follow.

There is a parallel NEPA process involving the preparation of environmental assessments, or EAs, which are prepared for projects that will have some impact on the environment but not so much as to require a full-blown EIS. Environmental assessments are generally far less complex than EISs. They need not include the public or interagency review requirements of EISs. Once an EA is prepared to the satisfaction of the agency, the agency prepares a FONSI. If it decides that the project will have a major impact on the environment, an EIS becomes necessary. However, such decisions occur rarely because the decision to prepare an EIS is usually determined by category of project, as reflected in agency rules implementing the CEQ regulations. Some major agencies have never made a decision that an EIS was necessary on the basis of an EA.

The NEPA process involves two important components, in addition to its requirements concerning EISs and related documents. One is that agencies must develop interdisciplinary staffs to manage their EIS process. Public participation is the other component. CEQ's (1970) requirement for public, interagency, and intergovernmental review provided legitimate entree to agency decision routines. After other forms of public participation became common in resources planning during the 1970s, CEQ (1978: §1506.6) formally incorporated such participation activities the NEPA process.

The NEPA process was not so formalized and standardized prior to the 1978 CEQ regulations. However, virtually the same processes were already occurring, under a wider range of names, within a few years of passage of NEPA.

THE FORMAT OF EISs

The contents and format of EISs have remained fairly stable since the early 1970s. Some standard sections of EISs stem directly from the statutory requirements of NEPA, while others are based on provisions of early CEQ guidelines that have been perpetuated. Each EIS naturally varies somewhat in the order and arrangement of the contents and in the amount of detail devoted to particular topics; in particular, EISs on complex or controversial projects tend to be longer than EISs on routine proposals. Table 2.1 uses the contents of the Forest Service's (1976a) final EIS on the Mineral King ski area to illustrate the organization of environmental impact statements. This long and thorough EIS dealt with a proposal that was a cause celebre of the Sierra Club and the subject of a landmark Supreme Court decision. The Forest Service's efforts were for naught, since the development was eventually abandoned in the face of continued opposition and complex, high-level Washington, D.C., politics. Nonetheless, the document provides a fine example of a well-structured EIS.

The first few pages of an EIS are the "summary sheet" required by CEQ guidelines. Despite the term "sheet," these summaries are invariably four to six pages long. The summary sheet contains standard classification information, summaries of the proposal and its major impacts, and a list of agencies that are to receive copies of the document. In other words, the section performs the same function as the executive summary commonly found in government reports.

The first text chapter in an EIS describes the proposed project, including its location, engineering or design specifications, background, and rationale. Chapters 1 and 2 of the Mineral King FEIS fulfill these functions. In this case, the Forest Service apparently felt that the proposal was so controversial that the rationale for the development merited a separate chapter. NEPA does not specifically mandate a description of the proposal, but such a descrip-

38

Table 2.1
Contents of an example environmental impact statement—the
Mineral King Ski Area Final EIS (Forest Service 1976a).

Chapter	Number of Pages
Summary Sheet	5
1. Introduction	
Purpose and scope of the environmental statement	2
Goals and objectives of the proposed action	2
Reasons for developing Mineral King	6
History of the proposal	2
2. Description of the Preferred Alternative	
Project location	6
Proposed recreation facilities and management	10
Preferred transportation system	1
Support systems	2
Land requirements and mineral withdrawal	2
Future plans	1
3. Environmental Setting Without the Project	
Physical environment (5 subsections)	38
Biological environment (3 subsections)	17
Cultural environment (11 subsections)	58
4. Environmental Impact Assessment of the Proposal	
with Mitigation Measures	
Physical environment (4 subsections)	25
Biological environment (3 subsections)	11
Cultural environment (11 subsections)	35
5. Adverse Impacts That Cannot Be Avoided	1
6. Relationship Between Short-Term Uses and the	
Maintenance of Long-Term Productivity	2
7. Irreversible, Irretrievable Commitment of Resources	1
8. Alternatives to the Preferred Alternative	
Introductory summary of impacts	6
Recreation alternatives (5 alternatives)	18
Transportation alternatives (8 alternatives)	7
9. Conclusion Analysis	4
10. Consultation with Others	10
11. Comments (62 letters reproduced of 2,150 received)	252
12. Bibliography	10
13. Other Appendices	67
Total Pages	601

tion is a logical prerequisite to a discussion of its impacts, so CEQ's (1971) guidelines stipulate that such a section should be included.

Similarly, a discussion of project-induced changes requires some knowledge of baseline conditions in the project area. CEQ's (1973) guidelines added a requirement for a description of the existing environment. This is usually found in the second text chapter in an EIS or is combined with the project-description chapter. Some material like this conventionally appeared even before the 1973 guidelines. Chapter 3 of the Mineral King EIS is longer than most baseline-environment chapters, but it addresses a particularly comprehensive set of conditions. The "physical environment" section, for example, contains subsections on geomorphology, geology, soils, climate and air quality, and hydrology.

On the other hand, some information in such sections seems to be included simply because the agency has it available. Usually the discussion is presented in an outline format and pedestrian prose. There are exceptions, however. In an EIS in the study's sample, the Bureau of Outdoor Recreation (1973: 12), over then-director Jim Watt's signature, rhapsodized about a wetlands whose addition to a state park was proposed for federal subsidy:

Spring brings a delicately hued sweep of shooting stars, yellow puccoon, yellow-eyed and blue-eyed grass, and golden alexander, with splashes of red Indian painted cup and puffs of white cotton grass. A few weeks later, spiderwort, black-eyed susans, and prairie phlox take over. In the late summer, golden blossoms of Coreopsis and purple spears of prairie gayfeather light up the prairie, and white-fringed orchids show nature's closeup artistry. Fall brings a grand finale of goldenrods, asters, tiny ladies tresses orchids, and grasses of gold, brown, red, and purple.

For better or worse few EIS passages read like this.

EIS writers present their predictions of the project's environmental impacts in the next EIS chapter. This is obviously the key analysis mandated by NEPA. Usually this chapter closely parallels the chapter on the project's "environmental setting." Thus, the Mineral King FEIS section on physiologic impacts is broken down into subsections on geology, soils, climate and air quality, and hydrology. The biological impacts section also parallels the topics in the existing-environment chapter, with subsections on vegetation, wildlife, and fish impacts. The cultural environment section, the largest in the chapter, contains subsections on archeological, health and safety, historic, landownership, land use, public service, recreation, socioeconomic, transportation, visual, and wilderness impacts. Each subsection also contains a discussion of measures to be taken to ameliorate any adverse impacts identified by the assessment.

The range of impact types discussed in the Mineral King document reflects the range of physical, biological, and social impacts found in EISs generally. This EIS does not, however, elevate economic impacts to the status of a major section. Economic and fiscal impacts take up the bulk of the Mineral King socioeconomic impacts subsection (along with miscellaneous social impacts, such as civil rights impacts). Most of the project objectives discussed in Chapter 1 are economic. Such an arrangement of economic predictions is fairly common in EISs. The documents tend to tone down any debates over alleged economic benefits by burying the former in misnamed "socioeconomic" sections or spreading economic justifications among various sections.

The next important chapter of the EIS covers alternatives to the proposed development, Chapter 8 in the Mineral King FEIS. Some statement of alternatives was a key feature of congressional intent in passing NEPA and was explicitly required by section 102(2)(C)(iii). Alternatives usually involve the relative size of the project and, under the CEQ (1973) guidelines, must include a "no project" alternative. The Mineral King FEIS presents alternatives for each of the two major issues surrounding the development. The first set of alternatives involve size, with a development range from a 20,000 people-at-one-time year-round resort to the status quo. The second set of alternatives are linked to a key legal issue in the litigation over the development access to the valley across Sequoia National Park land. The Mineral King transportation alternatives include two types of roads, low railway systems, bus transport, and helicopter flights. As this list indicates, EIS alternatives can be a bit fanciful. After describing alternatives, EISs mention the environmental impacts of each. However, the discussion of alternatives' impacts is usually less detailed than the proposal's impact discussion in the assessment chapter; the Mineral King example is fairly typical, with the description and impacts of 13 alternatives receiving less than half the space devoted to the impacts of the preferred alternative.

Three standard passages in pre-1979 EISs are treated as separate chapters simply because they are required by separate clauses--(ii), (iv), and (v)--in NEPA section 102(2)(C). The "adverse impacts which cannot be avoided," "short-term versus long-term," and "irreversible and irre-trievable" chapters invariably rehash points made in earlier chapters. The Mineral King FEIS's Chapters 5, 6, and 7 are typical of these redundant, one-page passages in EISs gener-ally--they say nothing interesting.

On the other hand, a part of a final EIS that is usually treated as an appendix contains some of the most important and interesting material in the document. CEQ's (1971) guidelines required both circulation of the DEIS for comments and a section discussing the problems identified during the public and interagency review of the DEIS. Thus, a final EIS contains a "consultation with others" section summarizing the distribution and comments received on the DEIS, a responses-to-comments section, and copies of letters commenting on the DEIS. If a small number of comments are received (i.e., 10-20 letters), usually all are reprinted at the end of the FEIS and EIS writers prepare individual re-sponses to the major points in each comment. If a large number of comments are received, the lead agency holds down printing costs by reproducing selected letters and providing one response to similar or repeated comments. In the Mineral King case, the Forest Service printed only what were deemed to be the most important 62 letters from among the 2,150 responses it received; these included letters from governmental agencies and the major interest group and institutional commenters. EIS editors regularly exclude products of letter-writing campaigns and similar low-thought inputs to the public participation hopper. The agencies also sometimes use content-analysis methods, pioneered by the Forest Service's CODINVOLVE program (Clark et al. 1974), to summarize large numbers of comments on controversial EISs. But this was unknown until 1973 and is still rare.

There are, of course, both random and systematic variations in EIS formats. There are some minor differences in format among federal lead agencies. Probably the most noteworthy is the Rural Electrification Administration's practice of requiring utilities to submit a large appli-cant's environmental report; REA then summarizes or repeats material from this report in its smaller EIS, officially incorporates the report by referring to it in the EIS's pre-face, and ships both documents out to all reviewers. Since implementation of CEQ's (1978) NEPA regulations, agencies have had the option to not reprint information that did not change between the draft and final EISs; thus, some post-1978 FEISs consist of little more than a summary sheet, a list of errata, copies of comment letters, and responses to comments. Many variations on the standard format are simply random differences; among the thousands of EISs printed during the 1970s, one can find unusual examples of any combination of project descriptions, existing environment, impacts of the proposal, and alternatives chapters.

CEQ's (1978) regulations standardized the mode of presentation that agencies had grown comfortable with, except that the three minor and redundant sections (e.g., Mineral King's Chapters 5-7) are no longer included. Thus, by and large, a final EIS produced in the mid-1980s has contents that are quite similar to those of EISs of the mid-1970s.

CONCLUSIONS

Internal agency requirements for certainty and standardization were important incentives leading to the development of fairly complex and elaborate routines for producing EISs. These procedures have led to EISs that are a little different, and in some ways a little more elaborate and broader in content, than seems to have been required by NEPA, as interpreted by the courts and the Council on Environmental Quality. But the external heat and noise created by the controversy and litigation over NEPA and its requirements in the early years of the act led to consistent agency procedures that seemed to expand the NEPA requirement to incorporate such elements as social and economic analyses. It also led to standard requirements to define the need for an EIS, to develop the standards and procedures for EIS preparation, to link EIS requirements with other agency requirements, and to create a common format for presenting the analysis.

In large part, the procedures developed in the early years of NEPA compliance continue to this day. Council on Environmental Quality guidlines, and particularly the 1978 regulations, were important in this standardizing process, but even the 1978 regulations basically amounted to codifying procedures that had already been adopted, in similar form, in many EIS-producing federal agencies. While many observers have linked the expansion of the NEPA mandate to court cases on EISs, the major evolutions discussed in this chapter were primarily the result of bureaucratic needs for certainty and predictability. The NEPA litigation was, at most, the obscure and indirect background noise against which this standardization occurred.

NOTES

1. The early NEPA precedents have been thoroughly reviewed by many analysts. See, for example, Anderson (1973), Liroff (1976: Ch. 5), and Wenner (1982). For exact legal citations, see the bibliography under Calvert Cliffs Coordinating Committee v. AEC, Committee for Nuclear Responsibility v. Seaborg, Environmental Defense Fund v. Corps of Engineers, NRDC v. Morton, Hanley v. Mitchell, SIPI v. AEC, Citizens to Preserve Overton Park v. Volpe, and SCRAP v. U.S.

2. Goldsmith and Banks (1983) provide a good review of the Supreme Court's restrictive rulings in environmental cases. For exact legal citations to these cases, see the bibliography under Scenic Rivers Association of Oklahoma v. Lynn, Kleppe v. Sierra Club, NRDC v. NRC, Sierra Club v. Morton [National Wildlife Refuge system], Catholic Action of Hawaii v. Brown, PANE v. NRC, and Trinity Episcopal School v. Romney. Also see the SCRAP II decision.

3
Project Implementation and Sample Selection

This chapter serves two key purposes. First, before we could select a sample of EISs for auditing forecasts, we needed to know whether the projects proposed by the EISs were actually implemented as planned. Second, we wanted to provide a unique study of implementation, in part replicating previous research (Culhane and Friesema 1977). For a random sample of 146 EISs, we catalog project types, completion status, and implementation pathologies, our term for the reasons behind implementation difficulties. The pathologies contrast political-institutional and rational-technical influences. We also provide detailed accounts of the field sample projects and their implementation. Thus, while this analysis served the practical purpose of sample selection, it contributes to the public policy literature by surveying implementation successes and failures and advancing a new perspective on the causes of implementation problems in a significant minority of our sample projects.

IMPLEMENTATION THEORIES AND CONCEPTS

At least since Pressman and Wildavsky's (1973) classic analysis of the Economic Development Administration's program to create jobs in Oakland, California, public policy scholars have endeavored to understand the critical phase of the policy process known as implementation. Public policies, after all, are rarely self-executing. As the literature evolved, great strides were made in recognizing that the process of public policymaking suggests a chain of causation. As Pressman and Wildavsky (1973: xv) explained, "Implementation may be viewed as a process of interaction

45

between the setting of goals and actions geared to achieving them. ... [It] is the ability to forge subsequent links in the causal chain so as to obtain the desired results."

Every public policy is formulated to produce a specific outcome. If the theory linking policy formulation to policy outputs is good, the desired outcome should be realized. But however good the theory behind a policy, desired outcomes do not happen unless the process of policy implementation is successful. Implementation entails putting policy theory into policy practice. In addition to their emphasis on these causal relationships in the policy process, Pressman and Wildavsky greatly expanded our understanding of the role of political behavior in meeting policy objectives, particularly with regard to what they refer to as the "complexity of joint action." It is this complexity that is often a contributing cause of implementation failure.

Despite the apparent interest in the study of implementation, most scholars still agree that we are only beginning to appreciate its importance to policy processes. As Mazmanian and Sabatier (1983: 4) point out, most theories about implementation are derived from the field of public administration or the systems approach to public policy. They suggest that we pay more attention to the match between stated objectives and policy outputs and outcomes, the modification of objectives, and the factors that affect implementation or alter the implementation process. In addition to such theoretical concerns, the literature is greatly in need of empirical implementation studies.

A useful definition of policy implementation is "those events and activities that occur after the issuing of authoritative public policy directives, which include both the effort to administer and the substantive impacts on people and events"; implementation is a dynamic process affected by a "web of direct and indirect political, economic, and social forces" (Mazmanian and Sabatier 1983: 4). At one time, implementation scholars tended to emphasize a simple model linking policy formulation to implementation. Many now seem to prefer the formulation-implementation-reformulation model in an express recognition of policy-making dynamics and the fact that the implementation process itself affects policy outcomes.

There are a variety of scenarios for how implementation takes place (Mazmanian and Sabatier 1983: 276-282). Success stories in policy implementation are harder to come by than accounts of failure. In the face of many impediments, the "effective implementation of major programs designed to appreciably alter the status quo is exceedingly difficult"

46

(Mazmanian and Sabatier 1983: 266). Mazmanian and Sabatier cite three fundamental sets of variables affecting policy implementation: the tractability of the problem, the statutory structuring of the implementation process, and the net effect of a variety of political forces on the level of support for a policy's statutory objectives.

Most public policy scholars identify a number of conditions as necessary, though not always sufficient, for successful implementation. For example, Robert Lineberry (1977: 77) points out that the timeliness of implementation can indicate a policy's success or failure: "If gearing up for a new policy is measured in years or even decades, implementation is not really successful." Implementation relies heavily on the enabling legislation that provides for a public policy. The probability of success is improved when this legislation poses clear policy objectives, a sound theory linking objectives to outputs and outcomes, and a game plan for implementation. Mazmanian and Sabatier (1983) provide a checklist for gauging the probability of successful implementation based on these considerations.

Even with the best possible enabling legislation, policy implementation depends on the skill of the implementing agency and political support from interest groups and governmental leaders. Over the long term, decisionmakers must maintain a sense of priority for the policy being implemented, particularly in the face of conflicts and changes that may weaken the theory behind the policy or its base of political support (Mazmanian and Nienaber 1979).

Much of the implementation literature deals with the implementation of governmental programs, which involves the execution of broad policy decisions embodied in statutes or comparable authoritative policy documents (e.g., precedential court decisions, major regulations, and executive orders). By comparison, project implementation means the execution of a decision that carries out a specific "major federal action." Almost all EISs are project-oriented, so some of the literature's concepts about implementation are not readily applicable. However, such valuable concepts as the complexity of joint action and the balance of constituency support seem relevant to implementation processes generally and the EIS process in particular.

The completion, defense, and approval of an environmental impact statement is far from the end of a policy process. EISs are not themselves governmental decisions; they are decision documents about proposed federal actions. At most, an EIS marks the end of the policy formulation phase and the onset of policy implementation. The art of environmental impact prediction is rendered moot without the implementation of proposed projects. Not every proposal that survives the EIS process comes to fruition. Nonimplementation occurs for a variety of reasons, although the complexity of joint action appears to be a leading cause (Culhane and Friesema 1977). If a project is not implemented, no one can be held accountable for the predictive capability of its EIS. For completed projects, the implementation process may affect the extent to which EIS predictions become reality. Indeed, as implementation scholars have pointed out, the process itself may lead to a reformulation of original policy goals.

Once a proposed project is implemented, resulting in a tangible policy output, predicted outcomes can be compared to actual outcomes. In other words, the assessment of environmental impacts depends on the actual implementation of proposed projects. The completion status of projects, therefore, was a guiding criterion for selecting the field research sample for this study. It was necessary to ascertain precise data on the status of the cases in the final sample because we needed a post-implementation period long enough to make reasonable measurements of environmental impacts for testing our principal research hypotheses.

SAMPLE SELECTION AND METHODOLOGY

The selection of the implementation study sample began with a computer-generated random drawing from the EISs catalogued in the Environmental Protection Agency's (1980) index of EISs. Four hundred entries were initially selected out of the cumulative total of 10,475 EISs filed. As reported in Table 3.1, 249 EISs were excluded from the analysis. These were draft EISs or supplements to draft EISs, very early (pre-1973) or very late (post-1977) EISs, EISs for projects outside of the lower continental United States, and generic EISs. This left 151 cases for analysis. Five more were eventually deleted from the sample because their completion status was unobtainable, leaving 146 cases in the implementation study sample. The sample selection logic is detailed in the methodology appendix.

After the exclusions, we sought to develop an implementation profile for each project. First, we wanted to ascertain the date of administrative implementation, which indicates when a project was first authorized and often when commitments of budgetary resources were made. For some projects, an administrative decision is the key implementa-

48

Table 3.1
Selection of the implementation study sample EISs.

Sample Selection Criteria	N
Random Sample	400
Sampling Error	
Blank entry	-6
Duplicate entry	-1
Draft EISs and Draft Supplemental EISs	-111
Date of EISs	
Pre-1973 EISs	-75
Post-1977 EISs	-36
EISs Outside Lower Continental United States	-11
Generic EISs	
Regional generics	-7
National or foreign generics	-2
Implementation Status Unobtainable	-5
Implementation Study Sample	146

tion threshold. Second, when physical construction was involved, we wanted to know when project construction commenced. Third, we wanted to find out when projects became operational, the date when they could be used for their intended purpose. Fourth, we wished to ascertain whether projects were delayed or modified and whether implementation was ultimately prevented as a result. Completion status could be gauged in part against explicit or implicit EIS schedules, as well as against the time frame generally expected for similar projects. Finally, we wanted to catalog the reasons for delays, modifications, and nonimplementation. Nonimplementation refers to those instances in which abnormally long and indefinite delays, suspensions, or cancellations occurred.

We collected implementation data using two instruments, a telephone survey and a mailed questionnaire. (For more information, see the methodology appendix.) Data were supplied by governmental officials and other persons responsible for or knowledgeable about the sample projects. Of course, the level of detail provided varied and the survey technique limited some data to perceptions and personal recall by informants. Our confidence in the data's accuracy on completion status is very high; it is naturally less so for the more qualitative assessment of implementation pathologies. Limited institutional memory or selective recall, for instance, could affect the quality of responses and some informants may not have been aware of the existence of an implementation impediment, such as a threatened lawsuit. For the purposes of the study, however, we found the pooled information from our two sources to be more than adequate for selecting the field sample and for an implementation analysis of the 146-project sample.

THE IMPLEMENTATION STUDY SAMPLE

The implementation study sample of 146 projects represents a fairly even distribution of EISs according to a number of characteristics. Geographically, the projects were found to be widely dispersed, located in 40 of the 48 lower continental states. To eventually facilitate selection of the field research sample, we organized the cases into geographic clusters ranging in size from as few as two to as many as ten cases within reasonable travel distances of each other for the purposes of the fieldwork. After project locations were mapped, 27 geographic clusters were identified. No particular geographic region of the nation seemed substantially underrepresented in the sample, although 22 cases were geographically isolated and thus were considered poor candidates for the field research sample.

The sample is also fairly representative of the range of EIS filing dates: 1972 (2 EISs), 1973 (29 EISs), 1974 (25 EISs), 1975 (28 EISs), 1976 (34 EISs), 1977 (20 EISs), and 1978 (8 EISs). Thus, the sample covers the period during which the NEPA process became institutionalized within the federal bureaucracy. It also provided ample choices for selecting the field sample in terms of identifying projects with post-implementation periods long enough to audit environmental impact predictions.

The distribution of the sample according to the lead agency for each statement is provided in Table 3.2. For comparison, the table also reports this distribution for all 10,475 EISs filed as of May 1980, the year our source was published. The table suggests that the distribution of cases in the implementation study sample is fairly represen-

Table 3.2

Implementation study sample by lead federal agency. Source:
Environmental Protection Agency (1980).

Lead Federal Agency	All EISs N	All EISs Percent	Study EISs N	Study EISs Percent
Federal Highway Administration (FHwA)	3,277	31.3	42	28.8
Army Corps of Engineers (CoE)	2,031	19.4	36	24.7
Forest Service (FS)	776	7.4	15	10.3
Dept. of Housing & Urban Development (HUD)	710	6.8	10	6.8
National Park Service (NPS)	170	1.6	6	4.1
Nuclear Regulatory Commission (NRC)	274	2.6	6	4.1
Soil Conservation Service (SCS)	371	3.5	6	4.1
Environmental Protection Agency (EPA)	311	3.0	5	3.4
Bureau of Land Management (BLM)	145	1.4	2	1.4
Federal Aviation Administration (FAA)	558	5.3	2	1.4
Postal Service (USPS)	9	.1	2	1.4
Rural Electrification Administration (REA)	106	1.0	2	1.4
Agricultural Research Service (ARS)	12	.1	1	.7
Bureau of Outdoor Recreation (BOR)	46	.4	1	.7
Bureau of Reclamation (BuRec)	159	1.5	1	.7
Coast Guard (CG)	54	.5	1	.7
Economic Development Administration (EDA)	46	.4	1	.7
Energy Research & Development Admin. (ERDA)	7	.1	1	.7
Federal Power Commission (FPC)	129	1.2	1	.7
General Services Administration (GSA)	165	1.6	1	.7
Interstate Commerce Commission (ICC)	40	.4	1	.7
National Oceanic & Atmospheric Admin. (NOAA)	156	1.5	1	.7
Tennessee Valley Authority (TVA)	48	.5	1	.7
Veterans Administration (VA)	28	.3	1	.7
All Other Federal Agencies	854	8.2	—	—
Total	10,475	100.0	146	100.0

tative. With 42 cases or 29% of the sample, the Federal
Highway Administration, which accounts for 31% of all EISs,
is only slightly underrepresented. In second place, with 36
cases or 25% of the sample, the Army Corps of Engineers is
slightly overrepresented because the Corps accounts for only
19% of all EISs. The other major agencies in the sample are

Table 3.3

Implementation study sample by project category and type.

Project Category & Type	N	Project Category & Type	N
Transportation		Public Land Management	
U.S. highway	31	Land use plan	6
Interstate highway	6	Timber management plan	5
U.S. highway bridge	4	Wilderness plan	4
Airport	2	Recreation development	4
Railroad	1	Misc. land management	3
		Herbicide/pesticide appl.	1
Water Resources Development			
Waterway operation & maint.	11	Energy Projects	
Small watershed	8	Nuclear plant permit	4
Multipurpose dam	6	Hydroelectric dam	2
Miscellaneous navigation	6	Transmission line	2
Flood control	5	Nuclear research & devel.	2
Harbor development	5	Nuclear plant operation	1
		Nuclear waste disposal	1
Urban Development & Buildings		Power plant cooling	1
Community development	6	Coal plant	1
Government building	5		
Sewage treatment	5		
Urban renewal	4	Total	146
Water supply	2		
Hospital	1		
Military base	1		

the Forest Service and the Department of Housing and Urban Development. The National Park Service, the Nuclear Regulatory Commission, and the Soil Conservation Service each have six cases in the sample and the Environmental Protection Agency has five. The remaining agencies are represented by one or two cases. The only agency that is significantly underrepresented is the Federal Aviation Administration, with 1.4% of the cases in the implementation study sample and 5.3% of all EISs.

Table 3.3 reports the distribution of the implementation study sample according to types of projects proposed in the EISs. Although different federal agencies are naturally

52

associated with each project type, the categories are based on substantive features: transportation, water resources development, urban development and buildings, public land management, and energy.

The largest type, with 44 cases (30.1%), falls under the heading of transportation. Transportation projects include both improvements to the existing mass transportation infrastructure and construction of new facilities. Highways are usually federal-state projects with EISs written for relatively short road segments. The Federal Highway Administration provides the funding for U.S. highways and is the official lead EIS agency. It has also nominally filed more EISs than any other federal bureau-- more than one third of all EISs (EPA 1980). State highway departments, however, are normally responsible for writing highway EISs as well as for planning and constructing these projects. Thirty-one transportation projects are U.S. highways and six are interstate highways; all are based on FHwA EISs. Four projects, also under the jurisdiction of the FHwA, pertained to bridge improvements on U.S. highways. Two airport projects, one based on a Federal Aviation Administration statement and the other based on a National Park Service statement, and one railroad project proposed by the ICC are also in the transportation category.

Forty-one projects (28.1%) pertain to water resources development, the second largest project category. These involve the maintenance of waterways and other forms of water resource improvement and development. The Corps of Engineers, which has written the second highest number of all EISs (EPA 1980), is frequently the lead federal agency. Waterway operation and maintenance (11 EISs), implemented by the Corps, and small watershed projects (8EISs), implemented mainly by the Soil Conservation Service, are typical water management concerns. Multipurpose dams, miscellaneous navigation concerns, flood control, and harbor development, each with five to six projects, are all Corps projects.

Urban development and buildings, the third project category, is represented by 24 cases (16.4%) and includes all infrastructural improvements or construction projects not in the transportation category. Like the transportation projects, however, implementation often involves a federal-state partnership. Six HUD-sponsored community development programs and four urban renewal projects, three by HUD and one by the Economic Development Administration, are represented. Five government buildings, two of which are post offices, are included in the urban development and buildings category. Sewage treatment (5 EISs) and water supply (2 EISs) are projects undertaken by the Environmental Protection Agency, the Agricultural Research Service, HUD, and the Corps. The hospital project was sponsored by the Veterans Administration and the military base was proposed by the Coast Guard.

Public land management, the fourth largest category with 23 cases (15.8%), pertains to the development, use, and protection of public lands. Federal land management agencies, as a standard administrative procedure, draft comprehensive management plans for the units in their jurisdiction. The category consists of various land use plans (6 EISs), timber management plans (5 EISs), and wilderness management plans (4 EISs) implemented principally by the Forest Service, the National Park Service, and the Bureau of Land Management. Four land management projects focused on recreational development, including a ski facility in a national forest and a state park expansion requiring land acquisition. The three miscellaneous EISs were two land exchanges involving public lands and a timber road. The herbicide application EIS was written by the Forest Service.

Finally, the smallest category includes the 14 energy projects in the sample (9.6%). A majority represent various phases of the nuclear use cycle: research, fuel processing, electricity generation, and waste disposal. Some are retrospective EISs, meaning they were drafted for completed projects. The Nuclear Regulatory Commission takes the lead in the category because of its authority over nuclear power plant construction (4 EISs) and operation (1 EIS) and nuclear research and development (2 EISs). Another research project was proposed by the Energy Research and Development Administration. Two hydroelectric plants are also in the sample, one sponsored by the Bureau of Reclamation and the other by the Federal Power Commission. The Federal Highway Administration and the Rural Electrification Administration each sponsored a transmission line project. The coal plant was another REA project. The nuclear waste disposal site was proposed by the Atomic Energy Commission and located at a national laboratory. The power plant cooling facility involved a statement by the Environmental Protection Agency.

This categorization is not exhaustive. It serves only as a heuristic device for organizing the sample projects for this study. Nor are the categories completely exclusive. Some projects could be placed into more than one category. Sewage treatment plants, for example, pertain to both water resources and urban development, but were placed in the latter category because they generally have more to do with infrastructural improvement than waterways. A timber road, which obviously involves transportation, is for equally obvious reasons included in the public land management category. The hydroelectric dams and the cooling facility touch on both water resource and energy concerns, but were placed in the energy category because they are fundamentally energy projects. The airport projects caused us particular difficulty. Although their inclusion in the transportation category is sensible, the structural nature of the projects and their impacts could also place them within the urban development and buildings category. Further, the Jackson airport EIS was prepared by the National Park Service because the facility is located in Grand Teton National Park. In the implementation study, this case is included in the transportation category. It was reclassified as a public land management project for the field research, including the implementation analysis of the field sample reported later in this chapter. Because of the nature of the environmental impacts at Jackson airport and the controversy surrounding them, we were persuaded to emphasize land use issues over air transportation for our investigation.

COMPLETION STATUS OF THE STUDY SAMPLE

Table 3.4 reports on the completion status of the sample projects by project category. Completion status is organized into four categories. First, for many types of projects, implementation involves a continuation of existing administrative activities. This includes waterway operation and maintenance, nuclear facilities relicensing, land use planning, timber management, and wilderness management. For most of the 29 projects in this category, project implementation can be equated with administrative approval of proposals. In other words, a legal authorization, a budgetary appropriation, or some other form of approval allowed activities already underway to continue without interruption. These projects do not generally entail physical construction in the same manner as the other projects in the sample. Second, completion can be measured by the operational status of projects. Sixty-seven projects in the study sample became operational sometime during the 1974-1982 time period, by which we mean that that physical construction was completed and the constructed facility was opened for its intended use. In the third completion status category are 19 projects that were underway at the time of data collection and five others that were scheduled for implementation. This category, project in progress, is generally associated

55

Table 3.4
Completion status of the study sample by project category. Numbers indicate the number of projects.

Implementation Status	Transportation	Water Devel.	Urban & Buildg.	Land Mgmt.	Energy	Total
Ongoing Projects						
Operation and maintenance	.	12	.	.	.	12
Nuclear relicensing	.	.	.	.	4	4
Land use planning	.	.	.	6	.	6
Timber management	.	.	.	4	.	4
Wilderness management	.	.	.	3	.	3
Operational Projects						
1974	.	2	.	.	.	2
1975	.	.	1	2	.	3
1976	3	1	2	1	1	8
1977	5	3	1	2	1	12
1978	2	1	1	.	.	4
1979	6	1	1	.	1	10
1980	6	1	4	1	1	12
1981	3	.	3	2	.	8
1982	5	2	1	.	.	8
Projects in Progress						
Underway	3	5	7	1	3	19
Scheduled	2	2	1	.	.	5
Projects Not Implemented						
Abnormal delay	3	4	.	.	.	7
Suspension	4	6	2	.	.	12
Cancellation	2	1	.	1	3	7
Total	**44**	**41**	**24**	**23**	**14**	**146**

with later EIS filings. Our informants for these five projects did not attribute their incomplete status to difficulties; the projects simply awaited implementation.

Finally, 26 projects, or 17.8% of the sample, were not implemented; the reasons for each are discussed below. Nonimplementation is not necessarily a permanent condition, but it does entail, at the very least, an indefinite postponement of the project. Cases of nonimplementation occurred across project types and lead agencies. The implementation of seven projects was discovered to be abnormally delayed. For these projects there appeared to be a general expectation that implementation would eventually take place, although there was considerable uncertainty about the length of postponement. Twelve projects were suspended after implementation had begun. Their future was also highly uncertain. Finally, seven projects in the sample were permanently canceled for a variety of reasons. More cases of nonimplementation were found in the water management category than in the others. However, the most severe and only permanent form of nonimplementation, cancellation, affected three energy projects, two of which involved nuclear power plants. The other canceled projects were a transmission line, a multipurpose dam, a railroad, a U.S. highway, and a land use plan.

We also attempted to measure project delays and modifications. Excluding the seven canceled projects, a total of 28 projects, or 20%, were delayed during implementation. In some cases delays were only minor postponements attributable to scheduling problems. In seven cases, however, delays evolved into the lengthy, abnormal delays described above. Modifications, some minor and some major, were reported for 30 cases (22%), again excluding the canceled projects. Minor modifications usually involved minor design changes that did not affect the overall character of the project. Major modifications entailed changes in project scope, usually reductions in scope, as when a nontrivial part of a proposal was dropped altogether. The problems of delay and modification seemed at times to be related; in 12 cases both occurred during implementation.

IMPLEMENTATION PATHOLOGIES

The coding scheme for the telephone survey and written questionnaire provided an open-ended approach for identifying reasons for implementation difficulty. Reasons varied widely, yet they fell into general categories of implementation impediments. We use the concept of pathology to describe implementation problems that interrupt the causal path in a policy system between policy formulation and policy output. Pathologies are symptoms of implementation difficulty that usually suggest the causes of difficulty as

57

well. Though often only one pathology is operative, it is not unusual to find multiple pathologies jointly inhibiting implementation. Some pathologies pertain only to temporary setbacks, as when minor scheduling changes are made. Others are immediately fatal to the project, as when a project is not authorized or is otherwise denied the resources necessary for implementation. Most fall in between, causing nonimplementation in some cases but not others.

According to our informants, 61 of the 146 EIS projects in the sample (41.8%) experienced some implementation difficulty, defined as a delay or modification in projects that have been implemented and defined as an abnormal delay, suspension, or cancellation in nonimplemented projects. Implementation problems, in the form of delay or modification, were reported for projects in each of the completion status categories: ongoing (8 EISs), operational (14 EISs), and in progress (13 EISs). Naturally, implementation problems were applicable in all 26 cases of nonimplementation. Problems were discernible in each of the five project categories: transportation (19 EISs), water resources development (19 EISs,), urban development and buildings (8 EISs), land management (9 EISs), and energy projects (6 EISs). For each of the 61 projects that experienced implementation problems, from one to five pathologies could be identified. Whenever informants indicated more than one pathology, the pathologies were not assigned ordinal values. In other words, no distinction is made between the first reason and the second, third, fourth, or fifth reason mentioned. Only one reason was offered for 25 EISs, two reasons for 15 EISs, three reasons for 13 EISs, four reasons for three EISs, and five reasons for five EISs. Seventeen implementation pathologies were discovered. Table 3.5 presents a typology of five pathology categories constructed on the basis of underlying patterns in the reasons for implementation difficulty.

The first category of implementation pathology, technical, means that implementation was inhibited more by technical features of the project and implementation process than by external forces. Technical reasons were mentioned 35 times. Admittedly, these pathologies are more descriptive than explanatory. In 19 cases a change in project scope was involved, which usually meant a reduction in project size. However, in at least two cases (a Corps operation and maintenance project and a Forest Service timber management plan) the authorization process permitted an increase in scope. Eliminating part of a project, such as a subsection of a road or a component of an urban renewal project, is a typical reduction in project scope. Reduc-

Table 3.5
Implementation pathology categories and types. Numbers indicate the frequency of mentions; problems were reported for 61 projects in the implementation study sample.

Pathology Category	N	Pathology Type	N
Technical	35	Changes in project scope	19
		Minor design modification	11
		Minor scheduling change/delay	5
Political/Legal	29	Litigation/injunction	11
		External agency decision/conflict	9
		Public opposition	9
Environmental	27	Need for study/supplemental EIS	12
		Siting considerations/conflict	8
		Environmental impacts generally	4
		Safety/quality assurance	3
Fiscal/Economic	26	Funding	17
		Project economics	7
		Market conditions	2
Institutional/ Procedural	14	Property acquisition/rights of way	7
		Governmental permits	4
		Legislative nonauthorization	2
		Jurisdictional change	1
Total	131		131

tions in scope were often accompanied by other pathologies, such as a loss of funding, an environmental controversy, or decisions by other governmental agencies. In 11 cases minor design modifications presented an impediment to implementation, even though they did not involve altering the overall scope or character of a project. Examples are a highway design change and changes in the particular amenities provided in a housing development. In five other cases

59

minor scheduling changes or delays, such as those caused by contractors, were blamed for problematic implementation.

A second category of pathology, consisting of 29 mentions, is political/legal. Political or legal implementation pathologies are manifestations of direct opposition to projects from external sources. In 11 cases litigation was initiated or threatened or an injunction was actually issued. The cases affected were four water development projects, four highway projects, one hydroelectric dam, one transmission line, and one airport. While the threat or initiation of litigation was usually a cause of delay, only two of the 11 disputed projects were canceled, while two others were suspended. The remaining seven were operational or underway at the time of this research. For nine other projects, decisions by or conflict with other federal agencies caused delay, modification, or nonimplementation. Four of these were Corps projects. Another nine projects generated public opposition that was substantial enough to thwart implementation at least temporarily. Four of these were highways and three were Corps projects.

The third pathology category consists of environmental issues. These were cited a total of 27 times and generally involved impacts on the natural environment and related conflicts. For 12 projects, the need for further study or better planning prior to implementation was an issue. In some of these cases a supplemental environmental impact statement was required and projects were delayed as a consequence. Siting considerations were mentioned eight times and usually entailed conflicts over the aesthetic quality of projects. For four projects, all of which involved Corps EISs, problems were attributed to a general perception of negative environmental impacts. Finally, safety and quality assurance matters affected implementation in three cases. These involved a multipurpose dam, a hydroelectric dam, and a nuclear construction project, all of which were delayed but eventually implemented upon assurances to the lead agency about quality and safety.

The fourth pathology category is fiscal/economic, which consists of 26 mentions. Funding problems, affecting 17 cases, were the most common impediment to implementation in the category. For most of these the issue was not simply the merit of projects, but their relative importance to the funding agency. More than half of those projects affected by funding considerations were highways, which often depend heavily on constrained and sometimes strained state budgets. A second form of fiscal/economic pathology has to do with the economics of projects themselves, or their cost effectiveness. For seven cases estimated costs outweighed perceived benefits, casting doubt on the long-term economic viability of the projects. Five of these were Corps water resources projects. Not surprisingly, the other two involved nuclear power plant construction, which is characteristically both expensive and controversial. Finally, two community development projects were affected by adverse market conditions, specifically sluggish local housing markets.

Finally, the institutional/procedural category includes the least frequently mentioned implementation pathologies. All 14 of those mentioned involved an external process that interfered with project implementation. In half of the projects difficulties were encountered in acquiring needed property or rights of way from outside entities. Four projects, two of which were Soil Conservation Service small watershed projects, likewise involved getting permits to ensure environmental quality. Permitting processes also delayed implementation of a recreation project and a transmission line. Legislative nonauthorization occurred for two cases, both of which were wilderness management plans that were implemented de facto. (These are two of the three wilderness management cases considered ongoing.) For a single case, involving the abandonment of a railroad route, nonimplementation was attributed solely to a jurisdictional change because in this case the authority of the lead agency, the Interstate Commerce Commission, ceased to exist.

The implementation pathology categories are compared with project categories in Table 3.6. The table reports only raw numbers, without controls for either the number of projects or the number of pathologies mentioned, so comparisons must be made with caution. In the transportation area, the political/legal and fiscal/economic pathologies prevail. In contrast, neither pathology category was mentioned for the public land management cases. For public land projects, technical reasons were most frequently mentioned. For the water resources development projects, the environmental pathology category was involved the most times, although technical and political/legal problems were also numerous. For urban development and buildings, implementation problems stemmed mostly from technical and fiscal/economic circumstances. Finally, there is a fairly even distribution across the pathologies for the energy projects, except for the singular institutional/procedural pathology.

Implementation pathologies are compared with project completion status in Table 3.7. For ongoing projects (i.e., operation and maintenance, nuclear facilities relicensing, land use planning, timber management, and wilderness manage-

Table 3.6

Implementation pathology categories by project categories. Numbers indicate the frequency of mentions; for each project, more than one pathology could be mentioned.

Pathology Category	Transportation	Water Devel.	Urban & Buildg.	Land Mgmt.	Energy	Total
Technical	8	11	5	8	3	35
Political/Legal	12	11	2	0	4	29
Environmental	4	15	3	2	3	27
Fiscal/Economic	11	7	5	0	3	26
Institutional/ Procedural	3	5	2	3	1	14
Total Mentions	38	49	17	13	14	131
N of Projects	44	41	24	23	14	146
N with Problems	19	19	8	9	6	61
(% of Projects)	(43)	(46)	(33)	(39)	(43)	(42)

ment), 13 reasons were given for implementation difficulty. Most of these (8 mentions) were technical, and fiscal/economic issues did not come into play at all. For implemented projects that became operational between 1974 and 1982, technical considerations again prevailed, accounting for 12 of the 32 reasons provided. Political/legal reasons, however, were mentioned nine times and other pathologies were also significant. For projects in progress, only those that were underway at the time of data collection reported implementation problems, and again technical reasons (10 mentions) were highly salient. Other pathologies, particularly fiscal/economic issues, also played a role.

For projects that were abnormally delayed, the 16 reasons for delay were spread evenly across the pathology categories, with political/legal and environmental reasons

Table 3.7
Implementation pathology categories by project completion status. Numbers indicate the frequency of mentions; for each project, more than one pathology could be mentioned.

Pathology Type	Ongoing	Opera-tional	In Prog.	Abnormal Delay	Suspen-sion	Cancel-lation	Total
Technical	8	12	10	2	3	0	35
Political/Legal	1	9	5	4	7	3	19
Environmental	2	4	7	4	7	3	27
Fiscal/Economic	0	4	5	3	10	4	26
Institutional/ Procedural	2	3	4	3	0	2	14
Total Mentions	13	32	31	16	27	12	131
N of Projects	29	67	24	7	12	7	146
N with Problems	8	14	13	7	12	7	61
(% of Projects)	(28)	(21)	(54)	(100)	(100)	(100)	(42)

leading with four mentions each. For suspended projects, informants placed blame on fiscal/economic reasons more than any other and did not mention institutional/procedural issues at all. The political/legal and environmental pathology categories also seemed to play a substantial role in project suspension. Finally, the 12 reasons cited for cancellation cover all the pathology categories except for technical. If any general pattern is discernible, it is that technical reasons may tend to be associated with the sort of project delays and modifications that do not necessarily result in nonimplementation. Nonimplementation, which usually involves multiple reinforcing pathologies, seems more likely to be blamed on fiscal/economic problems, followed closely by political/legal and environmental issues, than on the other categories of pathology.

The nature of implementation pathologies appears to be related to the eventual completion status of projects. Projects categorized as ongoing, operational, or in progress tended to experience implementation difficulty (delay or modification) for technical reasons. Evidently technical problems were overcome to the extent that the projects could actually be implemented. The findings are different for the cases of nonimplementation. When projects were abnormally delayed, suspended, or canceled, a fiscal/economic pathology was frequently involved. Thus, when implementation fails, lack of funding, inadequate project economics, or poor market conditions may be at fault. Fiscal complexities may be as salient as the general complexity of joint action.

THE FIELD SAMPLE

The implementation study sample of 146 projects provided the pool from which the field sample projects were drawn. It was desirable to have a field sample small enough to allow for in-depth analysis and large enough to allow for comparability and representativeness. Random distribution is the convention for sample selection in scientific research. For this study, sample randomness was sacrificed in favor of stratification. Thus, the field research sample's only connection to technical randomness is that it was drawn from the initial random sample.

The guiding criteria for selecting the field sample were the location of projects, timing of EIS filings, representativeness of lead agencies, comparability of project types, and project completion status. The sample selection process emphasized balancing research preferences within each criterion. Once the 27 geographic clusters of projects were identified, a step that removed 22 EISs from further consideration, the 124 remaining projects were organized according to project type: transportation, water resources development, urban development and buildings, public land management, and energy. Comparability of projects within these substantive categories was particularly relevant for methodological reasons. One of our key field research goals was to identify comparable impact predictions across more than one project (e.g., economic growth). This would allow us to collect and analyze comparable impact indicators (e.g., local employment statistics).

Once we knew the distribution of the sample according to location, lead agency, and project type, the EISs were ranked according to the filing dates of the final statements and the status of project completion. We preferred projects with post-1974 EISs and pre-1978 implementation as these would be optimal in terms of both the maturity of the EIS and the length of the post-implementation period for evaluating impacts. Later implementation dates and earlier filing dates made projects less attractive. Projects that were physically underway were considered on the basis of the proportion completed by 1978; if a substantial proportion of a "phased" project was complete by that time, certain impacts could still be reasonably assessed. Some nonimplemented projects were evaluated separately for possible inclusion in the sample for control purposes.

Our last step was to choose clusters of projects for the field travel. Whole clusters were added to or deleted from the sample in an attempt to satisfy our primary selection criteria. Thus, we chose clusters that would enhance the sample's comparability and representativeness while encompassing a variety of geographic regions, lead agencies, and project types. Within these constraints, we also preferred projects of significance over very minor projects. In sum, all the choices made in the sample selection process helped ensure that limited field research resources would reap the greatest return for the study.

The final field sample consists of 29 projects in ten geographic clusters. Case descriptions are provided below. Six clusters were within driving distance of our research base. The southern Lake Michigan cluster was considered the "local" cluster (2 EISs). The near-distance clusters were the Minnesota-Wisconsin (5 EISs), Eastern Ohio (3 EISs), Arkansas River (3 EISs), Smoky Mountains (2 EISs), and Virginia-Maryland clusters (3 EISs). The long-distance clusters, which required air travel by the field research team, were located in western Washington-Oregon (4 EISs), the Teton Mountains (2 EISs), North Florida (3 EISs), and Massachusetts (2 EISs).

The distribution of the field sample according to EIS filing dates is normal and the EISs represent the period after NEPA had become institutionalized. The modal EIS filing year was 1974 (9 EISs). Five projects were filed the year before, six the year after. One project was filed in 1972 and the remaining six were filed between 1976 and 1978.

The EISs were written by 13 lead agencies, another important indicator of their representativeness. More than one EIS came from each of the following: the Federal Highway Administration (6 EISs), the Corps of Engineers (5), the Atomic Energy Commission/Nuclear Regulatory Commission/ Energy Research and Development Administration (4), the

Forest Service (3), the Department of Housing and Urban Development (2), and the National Park Service (2). Seven other agencies each had one EIS in the field sample.

There are six cases within each project category, with the exception of energy, which has five cases. Each category is well represented. The transportation category is particularly cohesive, which is why we renamed it the highways category for the field sample. Its six FHwA projects are all related to road construction or improvement, although one involves a bridge and another involves an interchange. The water resources development category is dominated by five Corps projects, but also includes a Soil Conservation Service small watershed project. The urban development and buildings category is more diverse, including projects by four federal agencies. Two involve housing, two involve sewage treatment, one is an urban renewal project, and one is a government radar facility. In the land management category, two Forest Service EISs represent timber management plans and another involves a timber road. Two National Park Service EISs for Grand Teton National Park are also in this category. The sixth land management case is an acquisition and land use change involving the Bureau of Outdoor Recreation. Finally, the projects within the energy category are both diverse and comparable. They include four very distinct projects related to nuclear energy that were authorized by the AEC, the NRC, and the ERDA. A conventional coal plant, with an EIS by the Rural Electrification Administration, rounds out the sample.

PROJECT DESCRIPTIONS

One of our research goals was to have comparable projects within each project category to enhance our ability to generalize about environmental impact prediction. Yet every case in the field sample is unique, as revealed in the following accounts. Each brief case description indicates the project's nature as well as its implementation history.

Highways

State Road 24 Interchange, Gainesville, Florida. This project involves the construction of an interchange between I-75 and Florida State Road 24, upgrading of eight miles of State Road 24 between I-75 and the University of Florida campus to six lanes, and upgrading of an intersecting city

66

avenue to four lanes. The project's FEIS was filed in August 1973 and construction began in June 1974. The interchange and the State Road 24 segment were completed in March 1976, with the final segment completed one year later.

U.S. Route 53 and U.S. Route 8, Wisconsin. Litigation interrupted this project, which forms one link in the upgrading of U.S. 53 into a four-lane freeway. U.S. 53 is the major highway south from Duluth, and U.S. 8 is a major east-west highway through northern Wisconsin. Construction was underway when, in May 1973, the state was enjoined from further work on the project pending preparation of an EIS (Barta v. Brineger). Work on the U.S. 8 segment and the southern segment of U.S. 53 was completed in July 1974. The FEIS was filed in August 1974 and the northern segment of U.S. 53 was constructed between May 1976 and November 1977.

I-295, Jacksonville Beltway, Florida. This EIS deals with construction of the northwest quadrant of the I-295 beltway around Jacksonville, Florida, through a fairly undeveloped, thinly populated area. The project had been authorized in 1963 and the southwest quadrant around to I-10 had been completed between 1967 and 1973. The FEIS in the sample was filed in February 1974. Construction on the northwest portion of the beltway began in January 1974 and was completed in December 1978.

I-94 Interchange, St. Joseph, Michigan. This project upgraded the interchange between I-94 and Lake Shore Drive, the principal southern access to St. Joseph, Michigan. The interchange area is urbanized, with a moderate number of highway-related service businesses. The project was delayed for five years following the filing of an October 1973 FEIS. Construction began in April 1978, and the new interchange entered service in August 1979.

Skipanon River Bridge. This project is one of two in the sample that did not involve major action or significant impacts on the natural or social environment. The bridge is a 671-foot-long, two-lane replacement span on virtually the same alignment as the preexisting bridge. It crosses a small river that flows into the Columbia River estuary at the northwest tip of Oregon. The EIS was written because of a local controversy initiated by one vocal dredge operator who demanded construction of a drawbridge after learning that the existing bridge violated the terms of a 1916 Coast Guard permit. A drawbridge would permit dredges and boats with masts to sail upriver, where this particular operator's property was located. The FEIS was filed on the project in December 1975 and construction of a new fixed-span bridge began in December 1977. The bridge was opened in June 1979.

Wisconsin Highway 64, Dunn County. State Highway 64 lies about 15 miles north of Eau Claire, Wisconsin, and roughly parallels I-94. The project involved the reconstruction of the highway from Connorsville east to the Dunn County line. The FEIS was filed in October 1973. Construction began in June 1975, but because of its low funding priority, the project was not completed until August 1982.

Water Resources Development

Tacoma Harbor Channel Maintenance. The December 1975 FEIS on this project was an *ex post facto* document. The harbor consists of eight parallel waterways on the east side of Commencement Bay at the south end of Puget Sound. The March 1974 DEIS on the case proposed two distinct actions: dredging Blair Waterway, one of the main channels, to a depth of 40 feet, and extending a training wall on another channel, the Puyallup River. (A "training wall" trains the current so that it washes sediment out of the navigation channel.) A month after the DEIS was filed, acting under a special CEQ dispensation, the Corps let a contract for the dredging, which took place during June-September 1974. The training wall, representing 19% of total project costs, was deferred after the 1975 FEIS and action is still pending.

Cleveland Harbor Diked Disposal Site 12. This disposal site was one of several that received spoil material from ongoing maintenance dredging of Cleveland harbor in northeastern Ohio. Site 12 is located near the city's downtown lakefront and, once filled, was donated to the adjacent city-operated lakefront airport. The Corps' FEIS was filed in March 1973 and construction took place in 1974. The site began to receive dredging spoils in the fall of 1974 and was filled during the 1978-1979 dredging season. Although the Corps' diked disposal program is often controversial, this was a fairly typical project with routine implementation.

Paint Creek Dam. The Paint Creek FEIS, filed by the Corps in December 1974, was written on a "grandfathered" project begun, but not completed, before NEPA's passage. The project was part of the Ohio River flood-control program

68

first authorized in 1938. The Paint Creek dam, located west of Chillicothe in southcentral Ohio, was specifically authorized in September 1962 as a flood-control project with recreational and other side benefits. Reservoir land was acquired beginning in January 1966 and construction began in July 1967. The reservoir reached full elevation and three adjacent recreation areas were completed in June 1974. The Corps operates the dam, but the Ohio Department of Natural Resources manages the reservoir and adjacent recreation facilities as a state park.

Weymouth-Fore and Town Rivers Channel Rock Removal. This navigation project involved underwater blasting of rock outcroppings in the Weymouth-Fore and Town rivers, which share a channel outlet into Massachusetts Bay and form part of the Port of Boston. Rock blasting completed the deepening of the channel from 28 to 35 feet. The project's FEIS was filed in July 1974. Work began in September 1974, with the major blasting occurring from June 1975 to June 1976. Rock removal in the channel was completed by April 1977.

South Fourche Small Watershed Project. The Soil Conservation Service filed a March 1976 FEIS on this small watershed project, which was located in the Arkansas River basin west of Little Rock. The EIS proposed gradual implementation of seven flood-control structures and a flood-control and municipal water-supply structure. It also proposed conversion of 3,319 acres of floodplain behind the structures to grassland, conservation cropping and tillage on 6,904 acres, tree planting and timber stand improvement on 2,990 acres, and range management on 7,794 acres. The multipurpose structure, which provides water for the town of Perryville, was begun in June 1976 and completed in October 1977. Three flood-control structures were completed in March 1977, April 1978, and December 1980. One structure was underway, two were scheduled, and another was permanently deferred at the time of our research.

Cross-Florida Barge Canal Restudy. The proposal to build a canal connecting the Gulf of Mexico with the St. Johns River and Jacksonville, Florida, has been embroiled in environmental controversy for over forty years. Originally authorized in 1942, construction of the canal began in 1964. A 1971 presidential executive order suspended construction following a confrontation between the Corps of Engineers and conservationists led by Assistant Secretary of the Interior Nathaniel Reed. By that time, $75 million had been spent to construct all three of the project's dams, three of its five locks, and 25 of its 110 miles of channel work. A 1974 district court decision (E.D.F. v. Corps of Engineers) quashed President Nixon's impoundment of Cross-Florida Canal planning appropriations.

The Corps' August 1976 FEIS examined ways to dispose of the partially completed canal. The restudy was unusually complex because, in addition to considering various alternatives, it provided two recommendations. The "recommended completion alternative" involved a slightly modified alignment of the authorized canal. The "recommended noncompletion alternative" involved abandoning one major lock and preserving the remaining completed works, and required tangible engineering and maintenance actions. The chief of the Corps rejected completion, leaving noncompletion, with a slightly better benefit-cost ratio, as the default proposal.

In July 1978, after examining the restudy, the Carter administration recommended deauthorizing the canal and restoring its eastern Oklawaha River section. Despite legislative proposals from April 1979 onward, a final statutory decision about the Cross-Florida canal restudy has never been reached. Because the project lacks a conclusive disposition, its status is considered suspended for the purposes of this analysis. In reality, the Corps has more or less implemented the 1976 "noncompletion alternative." It operates and maintains the relevant segments of the canal and has abandoned Eureka Lock to thieves.

Urban Development and Buildings

West Cummington Radar Facility. This project consists of a radar antenna housed in a 38-foot-high white geodesic dome atop a 50-foot-high tower, along with a control building and related electronic equipment. Its location is West Cummington, a small community in the western Massachusetts Berkshires. The system was one of several improvements in the Federal Aviation Administration's air traffic control radar network. The FAA filed its FEIS on the facility in September 1974. Local opponents filed an unsuccessful NEPA suit against the project (Cummington Preservation Committee v. FAA1975). This litigation was not manifested as an implementation pathology. Construction began on schedule in April 1975, and the radar was operational by March 1976.

Yakima Central Business District Grant. This Economic Development Administration grant subsidized additional parking, sidewalk beautification, and traffic improvements within the central business district of Yakima, Washington. The purpose of the project was to enhance the competitive position of downtown retailers against suburban shopping malls by improving the central business district's ambiance and infrastructure. EDA's FEIS on the grant was filed in July 1976. Construction of the $5 million project began in the fall of 1976 and was completed in one year.

BARC Sewage Treatment Plants. The Beltsville Agricultural Research Center (BARC), located in the Maryland suburbs of Washington, D.C., is the Agricultural Research Service's major experimental research station. BARC's December 1974 FEIS proposed to modernize the center's two sewage treatment plants, whose effluents violated state water quality standards. Reconstruction began at BARC's east-side plant in February 1976. When the facility did not function as designed, BARC hired a new contractor, who completed the east-side plant in April 1983. Construction at BARC's west-side plant began in October 1976 and was completed in June 1977.

Newington Forest Subdivision. This housing development is located in Fairfax County, Virginia, a suburban area south of Washington, D.C. Though a private development, the project was subject to an EIS under Department of Housing and Urban Development regulations because of its large size. Construction began immediately after the September 1977 FEIS. The houses and a few garden apartments in the development, which totaled 1,808 residential units, were gradually built over a six-year period through August 1982.

Shepard Park Development. This EIS proposed seven high-rise or mid-rise residential buildings in a six-block tract on the Mississippi River Palisades in St. Paul, Minnesota. HUD wrote an EIS on the development because one of the high-rise buildings, a senior citizens' residence, was subsidized under the section 236 housing program. The Shepard Park FEIS was filed in October 1977 and construction began one month later. A luxury high-rise apartment building was completed in December 1978, the seniors' building was completed in November 1979, and a high-rise condominium building was completed in November 1980. One mid-rise was ultimately built as an office building and a planned recreation facility was replaced by a townhouse complex. Two of the high-rise buildings had been deferred at the time of our fieldwork because of a sluggish local real estate market.

71

Henrico County Wastewater Treatment Plant. This project involved a major new sewage treatment plant in eastern Henrico County, Virginia, which includes Richmond. In addition to the main sewage plant, the project included construction of a land-treatment sludge disposal site, sewer lines, wastewater pumping stations, and decommissioning of old systems. The EPA, which manages the federal sewage construction program, filed its FEIS on the system in April 1978. Construction of the system's interceptor sewer began in June 1979. Construction of the wastewater treatment plant was scheduled for 1983-1984, and the project was to be operational in 1986. Though only 12% complete at the time of our fieldwork, and coded as underway, the Henrico plant was our best candidate for a major sewage treatment project.

Public Land Management

Grand Teton National Park Master Plan. The Park Service's September 1975 plan on Grand Teton National Park, as with most federal comprehensive land management plans, embodied no major changes in management policy for the park. Grand Teton, which was also the locus of the controversial Jackson airport, is a major destination park immediately south of Yellowstone in western Wyoming. Although a land management plan covers the whole range of management issues for a public lands unit, the Grand Teton master plan emphasized park preservation through constraints on recreational uses. In this sense it was typical of Park Service planning and policy during the period. Various elements of the plan were implemented from its March 1976 approval date through the time of the fieldwork.

Hiwassee Unit Plan. This Forest Service land use plan was prepared for a planning unit on the Cherokee National Forest south of the Smoky Mountains National Park and Knoxville, Tennessee. The agency staff chose this plan for a test of its then-new planning process. The center of the unit is an appealing and only moderately used scenic river. The unit's timber harvest is fairly light, but still draws complaints from opponents of clearcutting, as is common throughout the region. The plan was approved soon after the June 1975 filing of the FEIS, and has been implemented over the subsequent ten-year plan period.

Ozone Unit Plan. This August 1976 FEIS on an Ozark National Forest planning unit in northwest Arkansas proposed straightforward multiple-use management of the unit's timber, oil and gas reserves, and limited range, wildlife, and recreational resources. The plan received routine administrative approval in the fall of 1976. However, the 1976 National Forest Management Act revised Forest Service planning procedures. Preexisting plans were to remain in force until replaced by plans prepared under the newly enacted procedures. Ozark Forest officers admitted they forgot about the Ozone plan during the controversy over the new planning process. Nonetheless, the unit would have been managed similarly if the plan had been consciously followed.

Weyerhaeuser Road. This June 1974 EIS is the second in the field sample that does not seem to meet the criteria for writing an EIS. The project was located in the intermingled area of federal land and private land in the Cascade Mountains east of Seattle, Washington. The proposal essentially involved a 600-foot road right-of-way across Snoqualmie National Forest land to provide access for logging operations by the Weyerhaeuser Company on its private timberland. Weyerhaeuser could have cut a road on its own land, but its clearcuts would have been within the "visual impact zone" paralleling Interstate 90. The right-of-way proposed by the Forest Service EIS allowed an environmentally preferable road design. The road was approved in October 1974 and constructed in 1975. Weyerhaeuser clearcut the section from 1976 to 1981. As the current ranger commented during a fieldwork interview, "I was kind of awestruck that we would do an EIS on a rinky-dink project like this."

Jackson Airport. The proposed enlargement of the Jackson, Wyoming, airport was almost as controversial as the Cross-Florida barge canal case. The National Park Service drafted the EIS on the proposal because the airport is located within the boundaries of Grand Teton National Park. The proposal would allow jet service to Jackson by extending the airport runway, constructing a new control tower, and installing other technical improvements. The proposal was supported by local aviation and commercial interests, but resisted by the Park Service. Following the March 1974 FEIS, Secretary of the Interior Rogers Morton deferred the runway extension, the new tower, and jet service. Installation of the approved technical improvements began in 1975 and was completed in 1976.

Eventually, technological improvements made jet service feasible without a runway extension, so the FAA filed a 1977 DEIS on B-737 jet service. The DEIS was revised and refiled in 1979, and after overcoming persistent Park Service opposition, the FAA filed its FEIS in November 1980. One month later the Secretary of Transportation approved B-737 jet service on a trial basis. The Sierra Club responded with litigation but failed to secure an injunction against jet service. At the same time, the State of Wyoming appropriated funds for a new control tower. The tower issue became moot when, in the wake of an air traffic controllers' strike, the FAA withdrew its offer to staff a new Jackson tower. Nonetheless, the first scheduled jet flight into Jackson landed in June 1981.

On the issue of the airport's operational status, Secretary of the Interior Cecil Andrus revised its special use permit in August 1979, holding the airport to be a "nonconforming use" within the park, and terminated the permit as of 1995. In October 1980 the Park Service countered the FAA's revised DEIS with draft regulations that would have effectively banned jets at Jackson. Congress responded with an appropriations rider in November 1980 that allowed the FAA to proceed. In November 1982, the Reagan administration's Secretary of the Interior James Watt reversed his predecessor's decision. Six months later Watt signed another revised Jackson airport permit terminating in 2033.

Illinois Beach State Park Expansion. This FEIS, filed by the Bureau of Outdoor Recreation in January 1973, covered the largest grant ever made to a state from the Land and Water Conservation Fund. Illinois had planned since January 1971 to double the size of Illinois Beach State Park, located on the Lake Michigan shore in the northeastern corner of the state. Almost all the vacant or erosion-threatened parcels designated for the expansion were acquired from 1972 to 1979. The master plan for the park's new unit was delayed until 1982. Main recreational facilities were not constructed until 1984, after our fieldwork. The size and nature of the EIS and the lead agency added diversity to our field sample, but the early EIS date and indistinct implementation period were shortcomings. The project's proximity to our home office, however, was a distinct advantage.

Energy Projects

Monticello Nuclear Station. The Monticello Station is located on the Mississippi River northwest of Minneapolis, Minnesota. The AEC granted Northern States Power Company a construction permit for the plant in June 1967 and a provisional operating license in September 1970. Its reactor went critical in December 1970 and a full-power operating license was granted in January 1971. Monticello was the subject of a precedential nuclear licensing decision. State environmental regulators, who were generally regarded as opponents of nuclear power, sought to impose state safety regulations on the plant. However, the decision in Northern States Power v. Minnesota (1971) held that the Atomic Energy Act preempted the states in the area of nuclear regulation. The AEC filed its Monticello FEIS, which reviewed the 1971 provisional operating license, in November 1972. It supposedly bore no specific relationship to the Northern States case. However, the provisional status of the license was not amended until the plant's routine relicensing in 1981.

Sequoyah Uranium Hexafluoride Plant. Uranium hexafluoride is manufactured as an intermediate step in the early part of the uranium fuel cycle. Kerr McGee's Sequoyah plant, located southeast of Muskogee in eastern Oklahoma, is one of two uranium hexafluoride plants in the U.S. Its product is shipped to gaseous diffusion plants that enrich the concentration of fissionable uranium for use in reactors and defense weaponry. (Uranium hexafluoride is the diffused gas.) The plant received its initial AEC license in February 1970. The February 1975 FEIS in our field sample covered the plant's relicensing. In contrast to the Monticello case, the Nuclear Regulatory Commission (AEC's successor) relicensed Sequoyah after the EIS process, in October 1977.

Shippingport Breeder Reactor. This project, sponsored by the Energy Research and Development Administration, involved the test operation of a light-water breeder reactor in a small, preexisting generating station located on the Ohio River in the heart of the depressed steel industry region of western Pennsylvania. The 72-megawatt plant is dwarfed by the adjacent Beaver Valley nuclear station and by the Mansfield fossil station, less than one mile upriver. President Eisenhower attended the 1957 opening ceremonies at Shippingport, the first operational nuclear plant of the "Atoms for Peace" program. The plant was managed under an arrangement between the Navy reactor office, Duquesne Power Company, and the government's nuclear contractors.

The light-water breeder reactor (LWBR) program began in 1967. In June 1974 the pressurized water reactor at the Shippingport station was shut down and installation of the LWBR core began. The AEC had to be forced through litigation (Calvert Cliffs Coordinating Committee v. AEC 1971) to implement NEPA in its licensing and research programs. In the same time period, the agency was reorganized to form ERDA and the Nuclear Regulatory Commission. In June 1976 the five-volume Shippingport EIS was filed. The LWBR installation was completed in August 1977 and full power was achieved one month later. As a test reactor, the LWBR was intended to operate only long enough to permit proof-of-breeding tests. Consequently, it was shut down in October 1982 to begin defueling and Shippingport became the nation's first decommissioned nuclear power plant.

Alma Unit Number 6. The Rural Electrification Administration subsidizes capital projects for rural electrical cooperatives. Alma's six units, on the Mississippi River in westcentral Wisconsin, provide over one half of the generating capacity of the Dairyland Power Cooperative. The sixth unit increased the generating capacity at Alma from 200 to 567 megawatts. The Alma Unit Number 6 FEIS was filed in May 1975. REA approved the project in June 1975 and construction began in August 1975. Unit 6 was operational in December 1979, although a transmission line, a secondary component of the proposal, was not completed until May 1981.

Oak Ridge National Laboratory Waste Facilities. The AEC filed its FEIS on this project in August 1974 and it was approved in November 1974. The project was not of the same scope as the other field sample energy projects. The EIS proposed replacing and improving the intermediate-level radioactive waste processing system at the Oak Ridge National Laboratory, west of Knoxville, Tennessee. The new facility consisted of two evaporators, eight new 50,000-gallon storage tanks, and related equipment. Construction began in October 1975. The new storage tanks, the most critical component of the system, were operational in 1979 and the entire facility was operational in August 1980.

COMPLETION STATUS OF THE FIELD SAMPLE

The completion status of the field sample projects, summarized in Table 3.8, is of clear importance to our research. Six projects were continuations of ongoing activities: three land use plans, two relicensing proceedings, and a project involving operation and maintenance. The majority, 19, became operational sometime between 1974 and 1982. Three projects--the small watershed, one of the community development efforts, and one of the wastewater facilities--were underway at the time of our research. In each case, however, the phased nature of implementation made it possible to assess certain impacts. Finally, the controversial Cross-Florida canal restudy remained suspended at the time of this research. However, since nonimplementation was the recommended alternative and the followed course of action, we could evaluate projected impacts based on this alternative. No "pure" cases involving nonimplementation were included in the field sample.

As Table 3.8 also indicates, several field sample projects were delayed or modified during implementation. Delay affected three of the six highway projects, although all were eventually completed. In the Wisconsin cases, Highway 64 was delayed for funding reasons and the U.S. 53-U.S.8 project was delayed by litigation and slightly modified. Delay also affected Jackson airport and Alma unit 6. Two of the three projects that were underway, both in the urban development and buildings category, had been delayed. Minor modifications played a role in the Jacksonville I-295 project and both timber management plans. Modifications in project scope affected implementation of the Tacoma harbor operation and maintenance effort, the Shepard Park development, and the Jackson airport.

Within each of the first four project categories is at least one instance of significant implementation difficulty. Among the highway projects, U.S. 53-U.S. 8 led to a early NEPA decision (Barta v. Brineger 1973) over the need to prepare an EIS. In the Jacksonville I-295 case, a lawsuit was brought by a single property owner who believed his property, according to one of our informants, "should remain as a passive, natural area." As a result, construction limits and schedules were changed, although these were not viewed as having an impact on the project's final scope.

In the water resources development area, problems were most dramatic in the Cross-Florida canal restudy, for which five implementation pathologies were identified: general environmental impacts, project economics, public opposition, need for further study, and litigation. A presidential order delayed the project, as did a court order citing environmental considerations. Litigation initiated by the Environmental Defense Fund resulted in an injunction to halt construction pending Congressional appropriations. Congress directed further study, focusing on the issue of project economics. The final administrative recommendation to not implement the project was based on "marginal economic considerations," although environmental opposition is usually cited as the principal reason for nonimplementation.

Table 3.8
Completion status of the field sample.

Project (Agency, EIS Date)	Agency	EIS	Status	Problems
Highways				
State Route 24 Interchange, FL	FHwA	1973	1977	—
U.S. 53 and U.S. 8, WI	FHwA	1974	1977	Delayed
I-295 Jacksonville Beltway, FL	FHwA	1974	1978	Del/mod
I-94 Interchange, MI	FHwA	1973	1979	—
Skipanon River Bridge, OR	FHwA	1975	1979	—
State Highway 64, WI	FHwA	1973	1982	Delayed
Water Resources Development				
Tacoma Harbor Operation & Maint., WA	CoE	1975	O&M	Modified
Cleveland Harbor Diked Disposal, OH	CoE	1973	1974	—
Paint Creek Dam, OH	CoE	1974	1974	—
Weymouth-Fore & Town Rock Removal, MA	CoE	1974	1977	—
SouthFourcheSmallWatershed, AR	SCS	1976	Underway	—
Cross-Florida Barge Canal, FL	CoE	1978	Suspended	Delayed
Urban Development & Buildings				
Cummington Radar Facility, MA	FAA	1974	1976	—
Yakima CDBG Development, WA	EDA	1976	1977	—
Beltsville ARS Sewage Treatment, MD	ARS	1974	1980	—
Newington Forest Subdivision, VA	HUD	1977	1982	—
Shepard Park Development, MN	HUD	1977	Underway	Del/mod
Henrico Co. Wastewater, VA	EPA	1978	Underway	Delayed
Public Land Management				
Grand Teton Natl. Park Master Plan, WY	NPS	1975	Land Plan	—
Hiwassee Unit Plan, TN	FS	1975	Land Plan	Modified
Ozone Unit Plan, AR	FS	1976	Land Plan	Modified
Weyerhaeuser Timber Road, WA	FS	1974	1975	—
Jackson Airport, WY	NPS	1974	1976	Del/mod
Illinois Beach State Park, IL	BOR	1973	1981	—
Energy Projects				
Monticello Nuclear Plant, MN	AEC	1972	Relicense	—
Sequoyah Uran. Hexafluoride Plant, OK	AEC	1975	Relicense	—
Shippingport Breeder Reactor, PA	ERDA	1976	1977	—
Alma Station Unit 6, WI	REA	1975	1979	Delayed
Oak Ridge Natl. Lab. Nuclear Waste, TN	AEC	1974	1980	—

In the urban development and buildings category, two projects experienced implementation difficulty. The Shepard Park development was both delayed and modified in scope. There were also land acquisition problems and opposition to the project from neighbors. Acquiring the necessary land proved more expensive than anticipated because of sellers' "holding out for more money," and the size of the development was reduced as a consequence. The Henrico Wastewater facility experienced scheduling delays because of funding difficulty and threatened litigation. This project was massive ($165 million), involving a substantial amount of federal support and threatened litigation over funding. A phased construction schedule was eventually required for the project's implementation.

The Jackson airport was the project in the public land management area that experienced an array of implementation pathologies: a change in scope, need for restudy, public opposition, conflict with other governmental agencies, and litigation. At the heart of this case is the issue, to some the principle, of operating a commercial airport within the boundaries of a national park. Opposition to the project came from the Environmental Protection Agency, the Department of the Interior, and environmental interest groups, the most vocal of which was the Sierra Club. Support, however, came from the Federal Aviation Administration, the Department of Transportation, airlines, and the local tourism industry. The National Park Service eventually ordered a transportation study to evaluate the proposal. Litigation was brought by the airport's board against the Park Service, after which implementation began. The reduction in scope, however, meant that three elements of the proposal—a runway extension, an air traffic control tower, and an improved lighting system—were not put into place. One of our informants concluded, "This is a complex issue, worthy of a graduate thesis." Commercial jet service at Jackson airport was eventually allowed on a two-year trial basis.

Our field sample did not include an energy project with significant implementation problems. The only apparent difficulty was the delay in completing Alma unit 6. The Monticello nuclear plant was highly controversial in its early history, but the EIS in our sample involved only the Nuclear Regulatory Commission's relicensing decision, which our informants viewed as routine. Any remaining controversy was not manifested by an implementation pathology in our study. Clearly the most controversial cases in the field sample were the Cross-Florida canal and the Jackson airport. For each, five implementation pathologies were identified.

CONCLUSIONS

As in an earlier study (Culhane and Friesema 1977), whose conclusions echoed Pressman and Wildavsky (1973), we observed that difficulties in implementation often have arational causes. Many implementation failures are related to the existence of overlapping authorities and the complexity of joint action. Some problems, such as procedural ones, are direct manifestations of complexity, while others, such as environmental conflicts, tend to be complicated or magnified by multiple decisionmaking authorities. We also know that the lack of funds for a project may simply indicate fiscal constraints and nothing more. The concept of pathology proved particularly useful for understanding these aspects of implementation.

Our findings suggest that pathologies stemming from rational-technical causes may occur with frequency but they can be overcome. Technical problems tend to result in delays or modifications, but they do not necessarily prevent implementation. In our analysis, technical reasons were not mentioned in any instances of cancellation. In contrast, institutional/procedural problems, which presumably are key symptoms of complexity, were the least frequently mentioned pathology. Yet these played a role in two cases of cancellation. Nonimplementation was more often attributed to fiscal/economic, political/legal, and environmental considerations. Unlike minor technical details, these factors are generally outside the control of agencies responsible for project implementation. Thus, they are not easily explained by rational decisionmaking models.

As we improve our understanding of implementation pathologies, we will better understand the success stories in the policy implementation literature. For a dozen projects in the implementation study sample, our informants identified the need for further study as a salient issue. The same message is naturally applicable here. If our sampling is roughly accurate, more than 40% of EIS projects encountered some trouble in the implementation process. A logical next step in this line of inquiry is to examine the ways agencies can overcome implementation pathologies. Our next step, however, is to hold these environmental impact statements accountable for their predictive accuracy, to see if their implementation resulted in expected consequences.

4
Impact Forecasts:
The Contents of EISs

Forecasts about the consequences of projects are the key contents of EISs as decision documents. The preliminary sections of an EIS--the descriptions of the project and the current environment--lay the groundwork for assessing project impacts. EISs' core chapters on the environmental impacts of the proposal and alternatives consist of sets of impact forecasts. The comments and responses-to-comments appendix in a final EIS reproduces the public and inter-agency debate over the EIS writers' forecasts in the preceding sections. In short, forecasts are the EIS's reason for existence as an "action-forcing" document; they are the passages that show agency officials have complied with NEPA's mandate to consider the impacts of proposals.

The characteristics of forecasts also bear directly on the correspondence between EIS practice and the rational model of decisionmaking. NEPA's legal guidance embodies key steps in the rational-comprehensive model, since EIS writers must consider all relevant impacts of all reasonable alternative actions an agency could take in a given case. According to the prescriptive literature on environmetal assessment, a good identification of the consequences of alternatives should be marked by a set of technically sophisticated forecasts. That is, if agency officials are to reach an optimizing decision, they must weigh the consequences of alternatives using some kind of calculus. However, an optimizing calculus is no better than the predictions of consequences with which it works. If predictions are imprecise, any attempt to calculate the optimum alternative will be mere guesswork.

In fact, EIS assessments usually fall far short of the ideal of the rational model, so far short that we chose to not use the term "prediction," which is common in the EIS literature (e.g., Canter 1977, Marcus 1979). The word "predict" means "to foretell with precision of calculation, knowledge, or shrewd inference from facts or experience." "Forecast," in contrast, suggests that "conjecture rather than real insight or knowledge is apt to be involved."[1] The contents of a typical EIS thus are more aptly called "forecasts."

THE IDEAL EIS PREDICTION

Much of the literature on the NEPA process places heavy responsibilities on environmental assessment staffs. Methodologists such as Canter (1978), Leistritz and Murdock (1981), and Rosenberg et al. (1981) prescribe a technically sophisticated environmental assessment process focusing on predictions of future conditions and changes relative to preproject baseline conditions. They instruct EIS writers to formulate clear predictions that are informed by state-of-the-art theories and methods from relevant disciplines.

From this prescriptive literature one can identify several characteristics of the ideal EIS prediction. (See, in particular, Warner and Preston 1973 and Canter 1977: 176-177.) Clearly an ideal prediction must be quantified. The prescriptive literature contains innumerable descriptions of methods for calculating impacts. Advocates of quantified predictions are influenced by the fact that state-of-the-art methods in most disciplines are quantitative. These methods usually involve nontrivial mathematics with multiple parameters, log or exponential functions, and so forth. Even so subjective a phenomenon as a scenic view can be quantified. Environmental assessment methodologists also know that disparate impacts can only be optimized at the final step in an assessment if those impacts have been quantified. (See Canter 1978: 207-218, regarding the techniques, in addition to cost-benefit analysis, that are available for optimizing sets of quantified predictions.)

Quantification implies three corollaries. In quantifying impacts, one must normally specify the units of measurement in which the change is denominated, the population or unit of analysis where the change is measured, and the time period during which the change will occur.

It is possible to quantify predictions without using standard technical units of measurement. Some EIS writers have used the Leopold et al. (1971) assessment matrix, in which the magnitude and importance of impacts are rated on

82

ten-point ordinal scales. However, state-of-the-art assessment methods usually compute impacts using appropriate measurement units, such as Jackson turbidity units, decibels, dollars, and the Shannon species diversity index.

If predictions are quantified using standard technical measurement units, the prediction logically must specify the population to be measured. The prescriptive literature tends to advocate assessing impacts at different levels of affected populations, and particularly assessing cumulative regional impacts as well as local impacts. Canter (1978), for example, advocates separate analyses of "microscale" (local) and "mesoscale" (regional) effects.

Similarly, one should logically expect a quantified prediction to explicitly state the duration of the effect. The distinction between short-term and long-term impacts is critical in environmental assessment methods, with those terms specifically mentioned in NEPA section 102(2)(C)(iv). In addition, the exact shape of a long-term postproject trend is critical to certain predictions, such as annual discounted entries in a cost-benefit computation. Is the effect predicted to be a step-level, linear, curvilinear, or some other form of temporal trend? In short, an ideal prediction should often take the form of a time series, beginning with baseline figures and continuing with representative postproject predicted values throughout the impact period.

Almost all impact assessment methods advocate explicit discussion of the significance of a predicted impact (cf. Canter 1978: 294-310). A ranking of impact significance may be used in lieu of quantified predictions in some subjective methods; for example, the importance of an impact is one of the two scales in the Leopold (1971) matrix method. In some more quantified methods, such as the University of Georgia normalized scaling procedure (Institute of Ecology 1971), significance indices are used as weights in aggregating predicted values across different impacts.

Finally, like any other decisionmaker, an EIS forecaster operates in a world in which the future cannot be known with certainty. Some project impacts may be intuitively obvious consequences of executing a project, but most impacts are affected by a number of causes in addition to the project. The effects of those other causes are unpredictable, so an EIS's predictions must be uncertain. In the theory of rational decisionmaking, the probability of occurrence of a consequence may be formally included as a parameter in a decisionmaker's calculus (von Neuman and Morgenstern 1944: 50). Thus, EIS writers should convey the likelihood that a predicted impact will occur, and a thoroughly rational prediction would precisely quantify the probability of occurrence of the impact.

In summary, the ideal EIS prediction is quantified using a technically appropriate unit of measurement, and clearly identifies the affected populations or resources measured and the time when the effect is to occur, preferably by presenting the prediction in the form of a time series. It should also explicitly state the significance of the impact and estimate the probability of its occurrence. Furthermore, many of the methodologies endeavor to translate predictions into quanta that can be manipulated by some algorithm to determine which alternative is optimum.

This idealized version of NEPA forecasting places heavy demands on environmental staff people. The matter of forecast certainty or probability, for example, has been debated by decision theorists for decades. Including a parameter for uncertainty in calculations of a decision's consequences solves none of the conceptual barriers to comprehensive decisionmaking. If it is difficult to know the future distribution of some decision's consequences, then it is doubly difficult to know both that distribution and the probability-density function of each consequence in the distribution (March and Simon 1958: 137-138, Cyert and March 1963: 10). In the same vein, economists argue whether to inflate the discount rate in a cost-benefit calculation to reflect risk in decision calculations (cf. Eckstein 1958: 86-90, Thompson 1980: 25). Similarly, there are serious questions about the desirability of quantifying environmental costs and benefits, relying on agency notions of beneficiality, broadening the scope of the locus of impacts, and most other tasks involved in an "ideal" prediction.

We do not advocate this idealized version of the NEPA process (cf. Friesema and Culhane 1976). However, the ideal is not a "strawman"--a model proposed simply so it can be picked apart. As described above, the leading U.S. EIS methodologists advocate quantified assessment. This ideal also seems universal. The authors of the so-called Dutch study of environmental assessment in eight countries (Environmental Resources Ltd. 1984) equate adequate prediction with the use of quantitative "techniques." The University of Aberdeen group concluded its study of environmental assessment in the United Kingdom by recommending "that predictions contain quantitative information, when possible on . . . what variable is to be subject to impact, . . . the magnitude of that impact, . . . the geographical extent of the impact, . . . the timescale of the impact, . . . the probability of the impact occurring, . . . the significance of the impact, . . . [and] how much certainty or confidence may be placed in the prediction" (Bisset 1984: 477). Beanlands and Duinker (1983: Ch. 7-8), in the most influential study of Canadian environmental assessment, consider a quantitative approach and rigorous analysis of impact significance and probability essential to their ecological framework of impact assessment. Duinker (1985) goes so far as to entitle a paper on the subject "Better Quantitative and Wrong, Than Qualitative and Untestable"!

The ideal of quantified predictions represents the essence of competent engineering and rational planning based on the scientific method, so the "best"--that is, most sophisticated, state-of-the-art, rational--prescriptive work instinctively moves toward the ideal prediction. It maintains a powerful, normative, professional hold on the technically trained people who engage in the business of environmental assessment and the scientist-researchers who consult with environmental managers and write about environmental assessment methods. In other words, the ideal has been consistently advocated by scientists, engineers, and economists in the prescriptive literature because they believe environmental assessment should be done in a state-of-the-art fashion.

REALITY: THE VAGUE EIS FORECAST

Most passages in the EISs in this study's sample fell far short of the ideal prediction described above. All forecasts in the 29 field-sample EISs were content analyzed. (The content analysis protocol and reliability are described in the methodology appendix.) A "forecast" is defined as any passage in the text or responses-to-comments sections of final EISs about a future consequence of the proposed action. The field-sample EISs contain 1,105 forecasts, an average of 38 per EIS.

The Subjects of EIS Forecasts

According to Barry Commoner (1971: 29), "Everything is connected to everything else." To the extent that EIS writers follow the logic of Commoner's aphorism, EISs may be expected to discuss almost anything and everything. Indeed, the field-sample EISs address 188 types of project impacts. Certain impact types are mentioned regularly in EISs. Some

TABLE 4.1

Types of forecast impacts in the 29 field sample EISs. The total number of impact types is 188; several impact types with few forecasts are combined below. N = 1,105 forecasts.

Impact Type	Frequency

Physiographic Impacts (N = 301 forecasts)
Air
    Criteria pollutant compounds (e.g., CO, NOx)   30
    Fugitive dust   12
    Construction, general dust and emissions   11
    Other noncriteria pollutant compounds   3
    Fog, mist   2
    Particulates   1
    Air quality, inspecific   4
Water
    Turbidity, siltation   30
    Chemical pollutants, nontoxic   10
    Biochemical oxygen demand, dissolved oxygen   6
    Thermal pollution   6
    Groundwater, chemical pollutants   4
    Eutrophication, other limnological processes   4
    Nutrients (P, N, etc.)   3
    Fecal coliform, bacteria   3
    Salt pollution   3
    Toxic (acute), lethal chemicals   2
    pH (acidity-alkalinity)   1
    Water quality, general/unspecified/other   20
    Groundwater, general/unspecified/other   10
Hydrology
    Water flow, surface waters   6
    Water circulation (within a body of water)   5
    Channel alignment or stream gradient   5
    Flooding (as physiographic event)   4
    Aquifer recharge, groundwater   4
    Hydrologic capacity of land area   3
    Water consumption   3
    Water yield, runoff   2
    Evapotranspiration cycle, effect on   1
Soil
    Erosion, soil loss   35
    Topsoil, conservation of   2

86

Soil quality (i.e., pathogens, trace metals)	2
Geology, impacts on	2
Topography, impacts on	1
Hazards	
Radiation, ambient (air or water)	15
Radiation exposure, human	14
Radiation waste/material storage/disposal	9
Nuclear plant system accident	6
Herbicide/pesticide exposure, flora/fauna	3
Herbicide/pesticide exposure, human	1
Fire risk, forest/range or rural	4
Fire risk, inhabited areas	2
Hazardous materials, transportation accidents	3
Toxic chemical, oil spills	2
Other	
Shading, solar exposure	1
Wind exposure	1

Biological Impacts (N = 167 forecasts)

Fauna	
Wildlife habitat	23
Wildlife populations	11
Wildlife species, diversity	2
Aquatic habitat	10
Fish populations	8
Fish species, diversity	4
Plankton, aquatic microorganisms	7
Benthic habitat	6
Benthic organisms, populations	2
Bird habitat	7
Bird populations	5
Habitat, wildlife/fish/birds (combined)	5
Populations, wildlife/fish/birds (combined)	6
Species diversity, wildlife/fish/birds (comb.)	1
Rare/endangered species, general	3
Pest species	3
Insects (not as pests)	1
Vegetation	
Natural vegetation	13
Trees, forests, general effects	19
Timber stand conversion	4
Silvicultural productivity	2
Logging slash disposal	2
Rare/endangered plants	3

[Table 4.1, continued]

Algae (not regarding eutrophication)	1
Ecosystems	
Aquatic ecosystem, general	7
Ecosystem type conversion	4
Wetlands	3
Wild area ecosystem preservation	2
Other	
Impacts on unspecified biota	3

Economic Impacts (N = 167 forecasts)

Employment	
Employment, direct	14
Employment, secondary/indirect effect	7
Income	
Income, population within area	6
Income, specific groups	3
Sectors	
Agriculture production, dollar value, yields	9
Timber, commercial sales/cut	8
Retail sales, revenues	7
Recreation, tourism business development	5
Electricity, consumption/demand	4
Realty sales, revenues	3
Electricity, output	3
Oil, natural gas production	1
Energy consumption/demand, general	1
Mining, production	1
Range use, productivity	1
Manufacturing (other) sales, revenues	2
Cargo shipping, waterborne	2
Passenger traffic, air/rail/ship	1
Development, commercial	11
Development, industrial	4
Economic dev't, regional or local, unspecified	15
Public Finance	
Property tax, assessments/receipts	18
Project costs	8
Payments in lieu of taxes (U.S. to local gov't)	3
Sales, income, or other taxes	2
Benefit/cost ratio, analysis	2
Project revenues (to government)	2
Other	
Property value change, adjacent land	7
Access, transportation cost change	3

Flood damage change	3
Accessibility to businesses	3
Consumer energy costs	2
Industrial water supply	1
Irrigation water provision	1
Engineering R&D advances	1
Producer market diversity	1
Private cost change, other	1
Other unusual economic impact	1

Social Impacts (N = 447 forecasts)
General

Land use	19
Population	10
Community cohesion	7
Flood plain risk	4
Demographics	3
Discrimination, age/race/sex	1

Public Services

Sewage, wastewater load	12
Solid waste load	12
Municipal water supply	10
Police and fire	8
Education	5
Public transit	4
Health care	3
Stormwater load	2
Social services	1
Public services, misc., other	12

Irritants

Noise	44
Safety, nonoccupational/public	7
Safety, occupational	6
Odor	4
Vibration (i.e., from blasting)	3
Crime	2
Litter	1

Amenities

Aesthetic, scenic effects	35
Hunting, fishing use	12
Recreational facilities/units	6
Impacts on resources from recreation use	5
Boating, canoeing, etc.	4
Open space preservation	3

Wilderness, preservation experience/use	1
Camping use	1
Recreation use, misc./various/other	16
Transportation	
Traffic congestion	14
Traffic volume, vehicular	12
Access (e.g., to a business or property)	10
Traffic safety/accidents, vehicular	9
Convenience of travelers	9
Air traffic safety	5
Pedestrian or bicycle safety, movement	5
Parking effects	3
Misc. effects on air/water/rail transportation	2
Traffic impacts on community, general	1
Cultural	
Historical sites (only)	15
Archeological/cultural areas (only)	12
Archeological/cultural and historic areas	9
Scientific research, facilitation of	2
Public education/study, enhancement of	2
Housing	
Displacement, residential	11
Displacement, commercial or industrial	4
Displacement, residential and comm'l/ind'l	6
Housing quality, maintenance	3
Housing stock, quantity	2
Second-home development, effects	2
Public Land	
Land or regional plan, relationship to	8
Park designation, boundaries	7
Wildlife refuge designation, boundaries	7
Public lands special use permit	2
Wilderness, wild river, etc., desig./boundary	1
Other	
Conversion of land to fixed project structure	12
Other land acquisition or leasing of rights	2
Collateral project, justification because of	2
Collision/impact property damage	2
Public health, coliform	1
Communications, impacts on	1
Compliance with government regulations	1

[Table 4.1, continued]

Mixed-Category Impacts (N = 23 forecasts)
"Environmental impacts," unspecified	11
Energy conservation	4
Socioeconomic welfare, unspecified	3
Depletion of nonrenewable resources, reserves	2
Mineral reserves, effects on	1
Community growth, general	1
Watershed rehabilitation	1
Total	1,105

of these are fairly common project consequences. Any EIS on a project involving construction disturbance of land will routinely refer to erosion, siltation of streams, fugitive dust, and the relatively minor pollutant emissions of construction vehicles. Erosion and stream siltation, as shown in Table 4.1, are two of the most commonly mentioned physiographic impacts. Almost all EISs mention impacts on wildlife. EIS discussions of habitat, population, and species impacts are often indistinguishably interrelated, and the set of 11 types of impacts on wildlife, fish, or bird habitat, populations, or species would, if combined, constitute the single largest type in the sample, with 82 forecasts. Such fauna forecasts account for fully half of all biological impact forecasts.

Some impacts are regularly mentioned because of well-established legal or administrative procedures. Under Executive Order 11593 and the Historic Preservation Act, for example, agencies must consult with cultural heritage officials (e.g., the federal Advisory Council on Historic Preservation and its state counterparts) about all projects constructed, funded, or licensed by the agency. The consultation identifies properties eligible for or listed on a national register and the need for further survey work. EIS writers document their compliance with E.O. 11593 procedures in the cultural impacts subsection of the EIS, resulting in a large set of 37 historic and archeological forecasts in the sample. Similarly, EIS writers commonly mention impacts governed by laws about noise pollution (44 forecasts), clean air (31 forecasts), or clean water (64 forecasts). The rituals of American federalism require federal officials to appear sensitive to the intergovernmental consequences of their proposals, so property taxes and "PILT" (payments in lieu of taxes from the U.S. to local governments for federal lands that may not be constitutionally assessed for property taxes) are mentioned frequently in EIS economic-impact sections.

An EIS on a relevant type of project may mention a certain class of impacts several times, with relevant variations.[2] Radiation impacts (44 forecasts) are naturally major topics in the four nuclear facility EISs in the sample. Displacement of residential or business property owners (21 forecasts) is a standard sensitive impact of highway proposals. Forest Service EISs account for many of the 27 silvicultural forecasts. In other words, agencies' standard procedures and professional skills or the controversial issues commonly surrounding certain types of projects naturally lead to various patterns of mentions.

Many other impact types are rarely mentioned in the sample EISs; a third of the impact types in Table 4.1 contain only one or two forecasts. Some of these infrequently mentioned impact types are rather bizarre. We are not sure how any federal project could affect "the geology" (by creating a new fault?). Unfortunately, some bizarre impact types are mentioned frequently. The most common "mixed-category" type of impact, at 11 forecasts, involves allusions to "environmental impacts" in which it is impossible to conclude what kinds of environmental impacts the EIS writer means.

Three principal types of forecasts are included in the coding. Three fourths of the forecasts deal with impacts; most of these are projected aftereffects of the proposed action, although 62 (5.2%) are incidental to the action itself (such as property-owner displacement, which must occur before the project can be executed). Project objectives account for 8.4% of forecasts, and are evenly divided between objectives that would be direct outcomes of a project (48 forecasts) and stated objectives that require some intermediate or contingent occurrence (45 forecasts). Because of the requirements of the study's primary research design, the content analysis also included mitigation measures, which account for a sixth of the cases coded. Mitigations are conceptually distinct from objectives and impact forecasts, however, since they involve promised future actions rather than predicted consequences. Most mitigations consist of amelioration measures, although two dozen promised monitoring or follow-up studies.

Several unique forecasts in the sample deal with objectives or mitigations. For example, the objective of

Table 4.2
Substantive categories and types of forecasts. Cell entries
are numbers of forecasts and percents of total.

Category	Impacts	Mitigations	Objectives	Totals
Physiographic	200 18.1%	90 8.1%	12 1.1%	302 27.3%
Biological	137 12.4%	24 2.2%	8 0.7%	169 15.3%
Economic	141 12.8%	0 0%	27 2.4%	168 15.2%
Social	340 30.8%	65 5.9%	42 3.8%	447 40.5%
Mixed	14 1.3%	1 0.1%	4 0.4%	19 1.7%
Totals	832 75.3%	180 16.3%	93 8.4%	1,105 100%

the Shippingport project was to conduct a full-scale engineering demonstration of light-water breeder reactor technology. The prize for the strangest forecast among the
1,105, however, belongs to a mitigation in the Weyerhaeuser
road EIS, which proposed to ameliorate the aesthetic impact
of a color contrast in rocks exposed by a road cut by
painting the rocks with an asphalt emulsion.

Perhaps the most surprising aspect of the impacts discussed in the sample EISs is the distribution among the four
substantive categories of forecasts. The basic distinction
among categories is that physiographic forecasts deal with
nonliving natural phenomena, biological forecasts with
nonhuman living organisms and their habitats, economic
forecasts with business and other money transactions, and
social forecasts with human noneconomic phenomena.
Depending on how holistic one wished to be, certain impact
types or individual forecasts could easily be placed easily

into more than one category. Hunting, for example, could be a social or a biological impact, depending on whether the forecast focused on the hunter or the game. Flooding is a hydrologic event and often causes economic property damage. SO_2 pollution is an atmospheric chemical phenomenon, caused in substantial part by utility firms' production byproducts, that affects both human health and aquatic habitats. In coding forecasts, we ignored all such holistic and perfectly true arguments, forcing as many forecasts as possible into one of the four main categories.

The distribution of forecasts among the four principal substantive categores is different from what one might intuitively expect about EISs. According to statutory intent and court interpretation, NEPA requires decision-makers to balance impacts on the "biosphere" or "ecosystem" with economic and technical advantages of proposed agency actions (NEPA §2, Calvert Cliffs Coord. Comm. v. AEC 1971). Thus, the common image of an EIS is of a document that weighs biological environmental costs against economic benefits. However, these two categories together account for less than a third of the forecasts in the 29 field-sample EISs. The modest numbers of forecasts in certain key impact types within these categories are also surprising. Project supporters routinely seem to boost a proposal by claiming it will "create jobs." Among all objectives, the economic are slightly overrepresented, at 29%, but only 21 forecasts in Table 4.1 deal with employment effects. More-over, the most holistic ecological impact types account for only 1.8% of all forecasts: seven "aquatic ecosystem," four "ecosystem type conversion" and "eutrophication/limnological processes," three "wetlands," and two "wild ecosystem preservation" forecasts.

Physiographic impacts are only slightly overrepresented, with 27% of all forecasts (i.e., about the one fourth one might expect with four categories). EIS writers, as noted above, tend to most faithfully mention impacts subject to non-NEPA regulations. These other legal requirements, notably those governing air and water pollution and radia-tion, fall within the physiographic category. Half of all mitigations involve physiographic impacts. The most common mitigations are mundane measures to minimize construction-related soil erosion.

Social impacts account for fully 40% of the forecasts in the sample EISs, and this seems odd for several reasons. Two sets of social impacts are subject to non-NEPA regula-tions, and thus likely topics for frequent mention by EIS writers, but noise and cultural impacts (80 forecasts total)

94

account for only half the gap between the social and the second-highest categories. Several important sets of project objectives also fall within the social category-- traffic movement and safety for highway EISs, various recreation uses for public lands management plans, and wastewater loads for the two sewage treatment plant EISs-- but it nonetheless runs counter to common understanding to find half again more social than economic objectives.

As noted above, the NEPA process is legally biased in favor of consideration of environmental (that is, biological and physiographic) impacts. The courts have divided on the question of whether socioeconomic impacts can be "significant" within the meaning of NEPA's clause "major federal actions significantly affecting the quality of the human environment." The courts in Hiram Clarke Civic Club v. Lynn (1973), Trinity Episcopal School Corp. v. Romney (1975), and Image of Greater San Antonio v. Brown (1978) held that social impacts were not sufficient to trigger NEPA's EIS requirement in the absence of significant impacts on the natural environment. However, in McDowell v. Schlessinger (1976), which involved a military base closing like that in Image of Greater San Antonio, and City of Rochester v. U.S. Postal Service (1976), which involved an urban project comparable to the Hiram Clarke and Trinity projects, the courts held that socioeconomic impacts were sufficient to require an EIS. CEQ's (1978: §1508.18) NEPA regulations adopted the restrictive line of precedents. However, once the EIS threshold is crossed, the NEPA regulations (§1508.6) state that social impacts must be discussed along with natural environment impacts. In any case, this legal quibbling over whether social impacts fall within the zone of interests NEPA was intended to protect would not lead one to expect the disproportionate numbers of social forecasts across all three columns in Table 4.2.

The number of forecasts in the social category may be an artifact of the content-analysis procedures. The line between social and economic impacts is a fine one, as suggested by the common use of "socioeconomic" in the EIS trade. Impact types such as land use, flood plain risk, the whole range of public service demands, and property owner displacement certainly have economic implications, even though they seem aptly classified within the social category under our coding definitions. Similarly, sewage and solid waste loadings are close to the border between the social and physiographic categories, but the "fire risk" physio- graphic impact types are just as close on the other side. We are also sensitive to the possibility that the research team, composed of social scientists (although one had substantial education in forestry), may have been more apt to recognize social forecasts than other forecast categories. Nonetheless, these considerations would only narrow the gulf between social forecasts and the second-place category. There are simply more social impacts and objectives forecast in EISs than one would expect from the literature on environmental assessment.

The Precision of Forecasts

The most disturbing flaws in EIS analyses become evident when the actual properties of forecasts in the study's sample are contrasted with the ideal characteristics of EIS predictions--quantification, clear measurement units, explicit statements of impact significance and probability of occurrence, and so forth.

The linchpin of a rationalist prediction is quantification. As indicated in Table 4.3, less than a quarter of the forecasts in the sample are quantified. Another tenth of the forecasts assert that "no impact" will occur, which could be interpreted as quantified since they imply zero change. (There is a difference, however, between no change in the value of some indicator and continuity in the rate of preproject change in that indicator; EIS forecasts rarely specify which a "no impact" forecast implies.)

Quantified forecasts usually provide only a single post-project value; 164 forecasts contain a single value for a single indicator, 51 cases forecast single values for two or more closely related indicators of the same impact, and another 17 forecast that the impact will fall within certain bounds (e.g., "between 4.5 and 8.6 ppm," "up to 90 dBA"). An ideal prediction should be specified in time-series form, but only 2% of the forecasts in the sample are in time-series form. Moreover, our classification of a "time series" is quite liberal. The coding protocol demanded only a baseline and two post-project values, so a highway EIS forecast of traffic volumes in, say, 1980 and 1995 (plus a baseline reading) would be fairly typical of the "time series" forecasts in the sample.

Almost two thirds of the forecasts in the field-sample EISs are verbal statements, with no quantified projections in either the text of the passage or relevant tables and charts. Of course, in some cases a verbal forecast seems perfectly sensible, even if it does not match the ideal of a rationalist prediction. Almost all mitigations are coded as

96

Table 4.3
Characteristics of forecasts that are closely related to the
model of an ideal EIS prediction: quantification, significance, and certainty. N = 1,105 forecasts in 29 field-
sample EISs.

Forecast Characteristics	Forecasts N	Percent
Quantification of Forecast		
Quantified		
Time-series	21	1.9%
Single-number postproject value	164	14.8%
Postproject values, multiple indicators	51	4.6%
Bounded-values forecast	17	1.5%
Percentages, re nominal classification	9	0.8%
"No impact" forecast	123	11.1%
Verbal, unquantified forecast	720	65.2%
	1,105	100%

. . . .

Significance of Forecast Impact		
"High" (or synonym, explicitly stated)	32.	2.9%
"Moderate" (or synonym, explicitly stated)	8	0.7%
"Insignificant" (or synonym, explicit)	285	25.8%
Quantified, without explicit significance	163	14.8%
Vague/ambiguous significance statement	78	7.1%
No explicit statement of significance	539	48.8%
	1,105	100%

. . . .

EIS's Certainty About Forecast Impact		
Quantified probability	1	0.1%
Certainty guaranteed by situation	74	6.7%
Impact conditional on intervening event	62	5.6%
Probability implied by key words "will," "will not," "very likely," etc.	641	58.0%
Possibility implied by key words "may," "could," "may not," etc.	318	28.8%
	1,105	100%

"verbal" since the passages describe actions to be undertaken. Mitigations could conceivably be quantified in a
nontrivial way, for example, the number of acres of wildlife
habitat bought to replace habitat lost to a Corps project (a
most important type of mitigation politically, but a type
that does not appear among this study's 180 mitigations),
but in most instances it would be misleading or absurd to
try to quantify a mitigation. It would also defy common
sense to quantify certain classes of impacts. The AEC staff
observed, in the Monticello EIS, that the plant's architec-
tural design was not aesthetically pleasing. Aesthetics can
be quantified, but it makes no sense to do so in such a
subjective (and fairly gratuitous) passage. One can
certainly overquantify, as the Forest Service (1974b: 118)
Mineral King DEIS did in describing noise impacts: "the
sounds of . . . wind . . . characteristically range from 10
to 15 dBA for the wind moving through the vegetation. . . ."
    A verbal passage can also display some strengths missing
from a quantified prediction. The writers of the Corps'
(1974b: 53) Paint Creek dam EIS note that a condemnation
price may not fully compensate a displaced property owner:

> [T]he reluctance of those required to move from their
> homes is very strong evidence that the land has a
> greater intrinsic value to its owners than the cash
> value they receive. . . . A piece of land or house may
> have sentimental values because of the memories it holds
> for the owner . . . [or] the neighbors or the neighbor-
> hood in which it is located. Personal contentment and
> happiness of owning property is a definite, although
> intangible, value to the owner. . . . Other intangible
> costs may exist for the owner who . . . places a sub-
> stantial value on the creek which ran through his front
> yard [or] the large trees outside his back door. . . .

    More often, verbal forecasts reflect vague, uninforma-
tive, or unclear thinking. Discussing impacts on nearby
residential areas, the Yakima central business district
redevelopment EIS forecast, "The noise and greater presence
of vehicles along those streets could diminish the quality
of residential amenities in those areas, though by gradual
and unquantifiable degrees" (EDA 1976: 126). If EIS writers
chose to deal with a concept as fuzzy as the quality of
residential life, they will rarely know how to quantify the
impact. On first reading Table 4.1, one might think an
impact on the evapotranspiration cycle as bizarre as an
impact on geology. The actual forecast deals with an impact

of the Ozone unit plan's proposed conversion of hardwood timber stands to softwood: "Clearcutting a stand affects the evapotranspiration cycle on the stand area for a few years" (Forest Service 1976b: 93). The context of this passage does not make clear whether it refers to different rates of water runoff from bare versus forested slopes. At a minumum, if all the trees are felled, then no vegetation remains to transpire water vapor into the atmosphere. The Newington subdivision EIS, also dealing with hydrologic impacts, proclaims:

> Ground water recharge will be reduced with a consequent decrease in water table as a result of the additional impervious surfaces within Newington forest [but] the development plan and use of cluster housing with massive areas of natural vegetation should minimize this negative impact (HUD 1977b: III.2).

In other words, roads and roofs stop rain from percolating through the soil, but since most of this very low density development is open space, the groundwater table probably will not be affected in the least.

As the above passages suggest, verbal forecasts usually do not refer to any common unit of measurement. Fewer than half the forecasts are couched in a unit of measurement other than some concept that simply restates the forecast impact (Table 4.4). Mitigations are special cases in which the measurement unit can be treated as a dichotomy (the mitigation is either done or not done). However, two thirds of the forecasts about objectives and impacts do not use measurement units. Only 29% of the forecasts employ a technical measurement unit and, as discussed below, these measures are not evenly distributed among the various forecast categories.

The typical forecast also falls short of several other characteristics of the ideal EIS prediction. Over half contain only vague or contradictory statements about the impact's significance. As shown in Table 4.3, another 15% lack an explicit statement of impact significance, but are at least quantified. Thus, readers have some basis on which to judge the impact, even if they are not alerted to threshold or critical values that might affect their judgment.

The forecasts fall even further from the goal of an explicit estimate of the likelihood that an impact will occur. Table 4.3 lists only one forecast that quantifies the probability of occurrence of the forecast impact.[3] A larger number of forecasts state that the occurrence of the

Table 4.4

Selected listing of measurement units used in six or more forecasts in the 29 field-sample EISs. Some comparable measurement-unit codes have been combined; for example, ug/m³ and mg/l share one code, and barrels of oil and MCF of natural gas share another. If two or more measurement-unit codes are combined on the same row below, the number of codes is shown in brackets. N = 1,105 forecasts and 61 measurement-unit codes.

Units of Measurement [Codes]	Number of Forecasts
Rads, rems, curies, or similar [2]	16
mg/l, ug/m³, ppm, or similar [2]	15
Tons, other gross weight measures	8
Other physical measurement units [13]	30
Biological measurement units [2]	7
Dollars	55
Employees	14
Other economic measurement units [10]	20
Buildings, residential units, owners [3]	16
Decibels	14
Recreation visitor-days	8
Average daily traffic	8
Designated public land units (i.e., numbers of)	7
Accident rates	6
Population	6
Other social measurement units [9]	20
Acres	42
Other mixed-category measurement units [6][a]	31
Mitigation: amelioration measures	138
Mitigation: monitoring, further studies [2]	24
No measurement unit	620

[a] Includes codes for frequency of occurrence of a defined event (12 forecasts) and time (2 forecasts).

impact depends on an intervening event that may be out of the control of the lead agency. In other cases, the impact is logically certain to occur if the EIS proposal is implemented. For example, if an EIS states that certain acres of open land will be taken in acquiring highway right-of-way (and if the highway is located as described in the EIS), then the forecast property conversion must occur before construction begins. Forecasts with reasonably explicit indications of the likelihood of occurrence account for only an eighth of all forecasts. In most cases a reader can only infer a forecast impact is possible or probable from the differences between key words like "could" or "may" and "will" or "will not." However, such inferences are inexact and other factors, such as the sensitivity of an impact or agency norms, affect EIS language. The Forest Service, for example, encourages a direct writing style, so its forecasts tend to use "will" more than "can" and "may."

The remaining coding characteristics are shown in Table 4.5. Two characteristics, the location and timing of the forecast impact, are related to forecast quantification in theory; that is, a quantified prediction must indicate where and when the change will occur. Verbal forecasts are not always perfectly clear about these points, but with four exceptions apiece it was possible to generally determine both the time and place of the impact. Three quarters of the impacts were forecast to be long-term, but that proportion misleadingly includes a common class of incidental impacts. For example, a statement that construction of a highway will cause "the loss of 0.5 acre in [a] bird sanctuary" (Florida DOT 1973: 16) refers to a fundamental and permanent change that occurs during the construction period. Such passages were generally coded as "permanent or long-term." (See the methodology appendix.) Three quarters of the forecasts also deal with impacts on populations or natural resources within the immediate project impact area, with the next largest set of forecasts dealing with impacts at the county or municipal level (e.g., property tax effects). In other words, EISs focus primarily on local consequences and rarely on broader, regional, or, in Canter's (1978) term, "mesoscale" impacts.

The final noteworthy characteristic shown in Table 4.5 is the adverseness-beneficiality of the forecast impact. Significantly more forecasts were coded "beneficial" (549 of the total of 1,105 forecasts) than "adverse," but the number of beneficial forecast impacts is inflated. Most "no impact" forecasts assert that an adverse potential impact will not occur. EIS writers mean such a passage to be read

Table 4.5
Beneficiality, population, timing, and salience of EIS forecasts. Mitigations' and objectives' beneficiality are excluded since they are almost always beneficial by definition. Four forecasts' referent populations are uncodable. N = 29 EISs.

Forecast Characteristic	Forecasts N	Percent
Beneficiality of Forecast Impact		
Beneficial	178	21.4%
"No impact" (beneficial, since not adverse)	111	13.4%
Neutral, explicitly stated	8	1.0%
Adverse	428	51.6%
Ambiguous	105	12.6%
	830	100%
Location of Affected Population/Resource		
Project site or adjacent area	855	77.7%
Nearby municipality, county	189	17.2%
Substate (e.g., multicounty) region	16	1.5%
State (incl. state adjacent area)	3	0.3%
Multistate region	4	0.4%
Industry sector or similar	34	3.1%
	1,101	100%
Time Frame of Forecast Impact		
Construction period	199	18.0%
Short term (within 1 year after construction)	39	3.5%
Permanent, indefinitely long-term	841	76.1%
Unpredictable, intermittent (e.g., accident)	22	2.0%
Ambiguous statement of time frame	4	0.4%
	1,105	100%
Salience of Forecast Impact		
High: significant systemic impact	189	17.1%
Moderate: common, reasonable tradeoff	825	74.7%
Minor forecast	88	8.0%
Implausible, trivial	3	0.3%
	1,105	100%

as a "beneficial" forecast, and the coding protocol treated it as such, even though the impact is, in effect, neutral. The coding protocol also defined measures that ameliorate an adverse impact as "beneficial," so logically almost all mitigations were coded as beneficial. Reasonable people may disagree with the utility of some project objectives, such as spending $1.5 million to rebuild the Skipanon River bridge to conform to a 50-year-old Coast Guard permit, but most objectives are also logically coded "beneficial." Thus, Table 4.5 deletes mitigations and objectives from its tabulation of adverse and benefical forecasts.

Even with these unusual cases deleted, the sample fore-casts contain a high proportion of beneficial impacts. Over half the forecast beneficial impacts involve assertions of economic benefits, which conforms to the common presumption that EISs writers want to emphasize economic advantages to offset projects' environmental disadvantages. EIS writers are often quite creative in describing the silver linings of their projects' clouds. Aesthetic impacts appear especially susceptible to creative puffery:

> Long-range aesthetic qualities of the area may be enhanced by the impact of the [Tacoma harbor dredging] project on Port productivity and activity, which are contributing factors in making the Port a stimulating environment for citizen pleasure. Many citizens and their children find the activity within the Port, the movement of vessels and the handling and trucking of cargo, to be interesting and enjoyable. Children are able to learn first hand about the fascinating world of trade (Corps of Engineers 1975: 54).

> The appearance of a [Amax Company stripmine in Campbell County, Wyo.] is a disagreeable sight to some people. The stripping shovels or draglines along with the coal storage silos can be seen on the horizon for miles. The trucks and other mining equipment create offensive activity on a peaceful landscape. On the other hand, other people consider the mine and its attendant activity an interesting addition to the landscape (U.S. Geological Survey 1975: 153).

Our all-time favorite EIS passage is the claim that a dam that would inundate a pine grove would eliminate "noxious emissions into the atmosphere" from conifers (Corps of Engineers 1971: 11). After Fish and Wildlife Service commenters noted that hydrocarbon-based smells are one attraction of walking in a pine woods, the Corps withdrew this assertion from the dam's final EIS.

One sees the same kind of puffery in discussions of the significance of impacts. As indicated in Table 4.3 above, forecasts usually do not explicitly characterize an impact's significance. When a forecast explicitly deals with the subject, however, it is seven times more likely to allege that the forecast impact is insignificant than moderately or highly significant. EIS writers can be just as creative in assessing an impact's significance as its beneficiality. The FAA (1974: 19), for example, argued that its large, white antenna dome on a mountain in a western Massachusetts forest "would not present a significantly unpleasant appearance" as one moved away from the dome because its image would become smaller. Such passages support our view that EISs are not rational analyses, but documents written to defend a lead agency's proposal in an adversarial process (Friesema and Culhane 1976).

PATTERNS OF FORECAST IMPRECISION

The principal characteristic of the forecasts in the sample is lack of precision. To measure this, we computed a simple index that counts the number of characteristics of forecasts that are not in accord with the traits of an ideal EIS prediction. (See Table 4.6.) Only 14% of the forecasts achieve a perfect score of zero on this index, with an average index score across all 1,105 forecasts of 2.40. This imprecision index also provides a means of comparing the quality of various forecasts.

Environmental Categories

The scores on this index provide further evidence of the deficiencies of biological forecasts in the sample EISs. As shown in Table 4.7, biological forecasts average 2.78, while the other three major categories all average 2.3. The small set of mixed-category forecasts have one of the worst mean index scores, 3.42, of any identifiable subset in the analysis. Recall that the majority of mixed-category forecasts refer to unspecified "environmental impacts," so they clearly merit poor scores on any vagueness index. These forecasts probably were intended to refer to biological impacts at least in part, but we cannot know for sure.

Table 4.6
Distribution of a forecast imprecision index that counts the non-ideal characteristics in each forecast. Each of the following adds one point to the index: (1) unquantified ("verbal") forecast, (2) no clear unit of measurement, (3) vague or no explicit statement of significance, (4) likelihood vague or based on possibility (e.g., "may") keywords, (5) uncodable referent population, (6) vague time frame, (7) unclear direction of impact, (8) vague forecast beneficiality, or (9) trivial salience.

	Index value							Total
	0	1	2	3	4	5	6	7	
Forecasts	153	177	240	241	210	72	11	1	1,105
Percent	14%	16%	22%	22%	19%	6%	1%	0%	100%

Mean = 2.40   Standard deviation = 1.52

Infrequent quantification is a common flaw of biological forecasts. (The index is influenced by a forecast's quantification. Quantification, per se, determines only one point of the index's maximum value of 9. However, a quantified forecast should clearly indicate a unit of measurement, impact location, time frame, and direction, and receive no worse than a "quantified-without-explicit-significance" code. The mean index score of quantified forecasts is 0.61 and verbal forecasts 3.26, which is in line with the logical relationship between quantification and other forecast characteristics.) Except for the small deviant mixed-category set, biological forecasts are the least quantified, at 19%, of the 832 impact forecasts in the sample.

The distribution of the minority of forecasts with clear units of measurement provides additional evidence of the unbalanced treatment of biological impacts. (Refer back to Table 4.4.) Economic and social forecasts more frequently use clear units of measurement than do forecasts about natural environmental impacts. The dollar is the most frequently used unit of measurement, since almost any economic impact can be denominated in money. (A few

Table 4.7

Average imprecision index scores of selected subgroups of EISs or forecasts. N = 29 EISs.

Selected EIS or Forecast Types	Mean Index Score	N
Forecast Category		
Physiographic	2.31	302
Biological	2.78[a]	169
Economic	2.29	168
Social	2.32	447
Other, mixed-category	3.42[a]	19
Lead Agencies		
AEC, NRC, and ERDA (4 EISs)	1.90[a]	135
Dept. Agriculture (4 bureaus, 6 EISs)	2.27	234
Dept. Housing & Urban Development (2 EISs)	2.27	100
Federal Highway Administration (6 EISs)	2.41	233
Corps of Engineers (5 EISs)	2.59	169
Dept. Interior (2 bureaus, 3 EISs)	2.93[a]	88
Project Type		
Energy projects (5 EISs)	2.00[a]	214
Water resources (6 EISs)	2.39	221
Highways (6 EISs)	2.41	233
Housing, urban & buildings (6 EISs)	2.56	260
Public lands management (6 EISs)	2.64	177
Comments and Controversy		
Most comment letters (6 high-quintile EISs, 24–41 comments/EIS)	2.72[a]	261
Fewest comment letters (6 low-quintile EISs, 5–8 comments/EIS)	2.25	121
Most controversial projects (6 EISs)[b]	2.38	216
Totals, all 29 EISs	2.40	1105

[a] Difference is statistically significant at 5% (Anova and t > 2 tests).
[b] Cross-Florida, Jackson, Monticello, U.S. 53 – U.S. 8, Cummington, and Shippingport.

measurement units listed in the social and economic categories, such as decibels or kilowatts, rely on hard physical measurement techniques, but this does not diminish the point about the imbalance in EISs' technical analysis.) Biological forecasts are particularly bereft of measurement units. The two measures listed in the biological row of Table 4.4 are wildlife population counts and species counts, which are primitive indicators of biological processes. In addition, the acres measure in the mixed-category set is used about equally in biological (chiefly wildlife habitat) and social forecasts. Thus, only about two dozen forecasts employ biological units of measurement. In short, biological measures in the sample EISs are few, infrequently used, and not particularly sophisticated.

There are few other significant interrelationships among the characteristics of individual forecasts. In particular, an apparent difference between forecasts for beneficial and adverse consequences seems to be an artifact of a minor coding convention. Actually, adverse and beneficial impact forecasts are about equally quantified.[4] The conventional wisdom is that EIS writers try to cover up adverse impacts with vague language--and we have provided some examples of such tactics above--but the content analysis does not provide consistent support for that notion.

Lead Agencies and Project Types

Individual EISs' average scores on the vagueness index vary from the best score of 1.58 for the NRC's (1975) Sequoyah EIS to the worst, 3.20 and 3.53, respectively, for the Forest Service's (1974a) Weyerhauser road EIS and the NPS's (1975) Grand Teton master plan. As indicated in Table 4.7, these best and worst EISs are quite representative, since public lands EISs have poor scores and energy and particularly nuclear EISs have good scores on our index.

AEC/NRC environmental assessments are generally prepared by special environmental staffs at research facilities such as the Oak Ridge and Argonne National Laboratories. These environments seem to be conducive to technical, thorough approaches to EIS writing. Whatever one may think of many nuclear proposals, the engineers and technical specialists on the AEC's and its successor agencies' staffs clearly quantify. They also, as we shall see in the next chapter, generally conduct better post-project monitoring of projects than do other federal agencies' staffs. Some decision theorists who accept the premise that rational decisionmaking is generally inefficient and impractical nonetheless admit that rationalism may be sensible in situations in which a bad decision would cause unbearably high damage (cf. Lustick 1980). Surely nuclear plant licensing involves such situations because of the high economic and health costs of a major nuclear incident. It is thus appropriate that the sample's nuclear project EISs come closest to the ideal of rational environmental assessment.

In the middle of the distribution, the two HUD documents impressed the research team as substantively good EISs, and they had a respectable 2.27 average on the index. The Corps, whose projects and EISs are often controversial, scored slightly better than average on the vagueness index. Finally, many observers regard FHwA highway EISs as among the least interesting products of the NEPA process. Nonetheless, the highway EISs in the sample lie exactly in the middle of the distribution of the imprecision index.

The public lands EISs are at the other end of the scale. Public lands planning documents, because they deal with possible agency actions over ten-year planning cycles, can appear ill-defined. However, while the Grand Teton plan is rated as the worst in the sample, the two Forest Service plans (Ozone, 2.04, and Hiwassee, 2.28) are better than average. The Forest Service's planning goals (along with those of its departmental sister agency, the SCS) are generally precisely quantified—for example, "sell 4512 MBF of timber annually," "construct and maintain 25.9 miles of foot trails" (Forest Service 1975: 17-18). Such multiple-use goals, which result from a "management-by-objectives" approach to planning, naturally deflate the service's vagueness index scores. Simply, the Grand Teton EIS pulled down the average of the Interior EISs, and the Grand Teton and smaller Weyerhaeuser road EISs depressed the public land EISs' average.

EIS Controversy and "External Reform"

As noted in Chapter 1, a principal alternative to the rationalist model of the NEPA process is the "external reform" thesis. This argument holds that one is bound to be disappointed by the technical quality of EISs, but the NEPA process nonetheless provides a valuable mechanism for opening agency decisionmaking to the light of public and interagency scrutiny (Friesema and Culhane 1976). The pivotal feature of the NEPA process as a source of external reform of agency decisionmaking is the requirement that agencies provide a period for public, interagency, and intergovernmental comment on a draft EIS, and then respond to those comments in the final EIS.

In previous studies we have found that the number of comment letters received by a agency on a draft EIS is a useful indicator of the pressure directed at the lead agency during the NEPA review process (Culhane and Friesema 1977; Culhane, Armentano, and Friesema 1985). It is not a perfect indicator, of course, because all letters are not equal. A handwritten note from a local resident and a detailed, litigation-threatening critique from the Sierra Club Legal Defense Fund each counts as one letter, but the two surely have different impacts on agency staffers. In one of these studies, a case study of the scientific quality of power plant EISs' coverage of acid deposition impacts, we found differences in the statistcal effects of two comment-letter variables. The number of letters on the EIS that commented specifically on acid rain, which is an indicator of focused, "peer review" influences, was the strongest predictor of coverage of acid rain impacts. However, the total number of letters, which indicates the general level of public controversy about the proposal, was not strongly correlated with good coverage of acid rain (Culhane, Armentano, and Friesema 1985). This present study of the technical quality of EIS forecasts could not manageably distinguish between focused peer review and diffuse public controversy because of the wide variety of impacts evaluated. The comment letters contain all comments on the impacts mentioned in an EIS, but the total number of such letters is only a rough indicator of the technical peer review of an EIS.

In any case, the relationship between public comments and the vagueness index differs from that found in our prior studies. The 29 sample EISs received a reasonable range of public and interagency commentary, from a low of five comment letters on the Hiwassee and BARC EISs to a high of 41 letters on the Grand Teton master plan, with an average 16.6 letters per EIS. The number of comment letters is moderately related to the imprecision index, but the correlation, $r = .38$, has the wrong sign (recall that the index increases as forecasts become less precise) and the relationship is nonlinear. Over half the cases are scattered and uncorrelated in the few-comments, low-medium-vagueness area of a bivariate plot. For example, the six EISs with the fewest comment letters have a low-average 2.25 mean vagueness score; these are all fairly short EISs, with an average of only 20.2 forecasts per EIS (compared to the

109

sample average of 38.1). Then a dozen cases move in a broad arc with more comment letters and decreasing vagueness scores, as the peer-review thesis would suggest. Fully .20 of the .38 correlation is accounted for by a single extreme case, the Grand Teton EIS, with the most comments and the worst vagueness index.[5]

The various subsets of comment letters are also not related to forecast precision in the sample. There is far less variance in the number of comment letters per EIS from governmental officials than there is in the amount of detail in such letters. Many federal-agency letters consist of two paragraphs amounting to a courtesy "no comment" on the draft EIS. Other letters are much more thorough. EPA prepares detailed comments on all EISs (except its own) as a statutorily mandated standard procedure. The Department of the Interior also routinely comments systematically on EISs, and its letter on the Cross-Florida barge canal restudy, which concluded dozens of pages of criticism with a referral of the case to CEQ, was the weightiest comment letter in the sample. But three quarters of the EISs contain roughly a dozen routine government comment letters. Most of the sample EISs do not contain extensive environmentalist or business commentary, averaging only 1.4 and 1.1 letters, respectively. Only five EISs contain a record of a classic environmmentalist-versus-industry debate: the Cross-Florida canal, Shippingport breeder reactor, Grand Teton master plan, Jackson airport, and Weyerhaeuser timber road. The numbers of government, environmentalist, and economic-interest comment letters are not significantly correlated with the vagueness index.

Finally, we separately inspected the half-dozen cases that met general criteria for controversiality, apart from simply above-average numbers of comment letters. The Cross-Florida canal restudy is the most famous, nationally controversial case in the sample, and the subject of both litigation (EDF v. Corps) and one of the few official referrals to CEQ of an interagency conflict in NEPA's history. The Jackson airport controversy involved high-level decisions by Interior secretaries Morton, Andrus, and Watt, a subsequent 1980 EIS by the FAA, Sierra Club litigation against both the Interior and Transportation departments,[6] and a running interagency dispute between the Park Service and FAA that also culminated in an official referral of the 1980 EIS (CEQ 1985: 551-553). The Monticello nuclear plant was the subject of a precedential decision on federal preemption of state regulation of nuclear plant licensing (Northern States Power v. Minnesota). The Shippingport test reactor was part of the controversial program to develop a commercial breeder reactor, and its EIS record contained the most polarized debate among national environmentalist (10 letters) and industry (7 letters) groups and actors in the sample and the second highest number of comment letters (34) in the sample. Finally, two other cases, the Cummington radar facility and the U.S. 53 - U.S. 8 highway project, were the subjects of litigation (<u>Cummington Preservation Committee v. FAA</u>, <u>Barta v. Brinegar</u>), although the suits were local and nonprecedential. Controversy did not seem to affect these EISs forecast precision; their pooled index scores of 2.38 are virtually identical to the sample mean of 2.40.[7]

We do not wish to overstate what appears to be one unusual relationship in a small sample of 29 EISs. Many EISs in the sample received minimal public and interagency review, and their forecasts are randomly vague. The rest of the EISs were subjected to greater scrutiny and, except for the deviant Grand Teton case, their vagueness indices appear to improve with more review. Appearances, however, can be deceiving. For example, the Cross-Florida barge canal restudy received the fourth-highest number of comment letters and, as indicated above, these letters were quite detailed; this EIS also has a better-than-average vagueness index of 2.22. However, the research team regarded this EIS as the most obtuse document in the sample. The FEIS consists largely of multipage tables comparing the very complex set of alternatives evaluated by the Corps to resolve this aborted project. A reader cannot easily grasp what is and is not a forecast in this document. Most of the forecasts contain numbers, are coded as quantified, and thus have low index scores, but the passages are still substantively unintelligible. In other words, this sample of EISs does not substantiate the proposition that public and interagency review improves the technical quality of EIS forecasts, at least as "quality" is indicated by the ideal characteristics measured in this study's index.

DISCUSSION

The chief conclusion to be drawn from this content analysis is that most forecasts in EISs are imprecise in one way or another, and do not remotely resemble the ideal environmental prediction of the prescriptive literature. The most common forecast in the sample is an unquantified assertion, not couched in any commonly recognizable unit of

measurement, and lacking any clear statement of the significance or likelihood of the impact.

These findings confirm those of some previous studies.
The "vagueness index" is not a perfect indicator of assessment quality, since it must measure general characteristics
across many types of forecasts and impacts. Nonetheless,
across all types of impacts in a range of types of EISs,
only 14% of the sample forecasts achieved a perfect score on
the imprecision index. We found about the same level of
inadequacy in an earlier study that defined the technical
quality of EIS passages based on the state-of-the-art
scientific literature but examined only one type of impact
in one kind of EIS. Even allowing one-year grace periods
after leniently late milestone dates, EIS forecasts covered
state-of-the-art findings about acid deposition in only a
quarter of the eligible cases (Culhane, Armentano, and
Friesema 1985). Across the Atlantic, the Aberdeen study
found that the forecasts in British EISs "were often
expressed in vague, imprecise and 'woolly' language" (Bisset
1984: 468), and the Dutch sampling of 140 EISs from western
European and British Commonwealth countries and the U.S.
found an infrequent use of predictive "techniques," with
only 25% employing quantitative models and another 15%
relying on simple inventories.[8] It seems that, across a
variety of countries and methods and standards of evaluation, only about a quarter of EISs' forecasts pass muster.

EIS coverage of biological impacts appears to be the
weakest link in environmmental assessment. Biological
impacts are the least frequently discussed consequences in
the sample EISs, are least often quantified, rarely employ
sophisticated measurement units, and have the worst
vagueness index of the four main categories. Unsophis-
ticated biological forecasts are understandable in one
respect: there are relatively few analytical biological
measures per se. Many physical measures serve as indicators
of changes in ecosystems that will affect the living
creatures in those systems--pH of lakes, ppm of certain
toxic compounds, and so forth. But EIS writers rarely draw
out the biological implications of physiographic impacts
and, at least in the sample EISs, make little or no use of
those biological indicators that are available, such as the
Shannon species diversity index.

The NEPA process was influenced by the holistic
philosophy of some visionaries from the conservation
movement, including the movement's scientific wing in the
discipline of ecological biology (plus a Congress dissatisfied with agencies' decision procedures). NEPA came to pass

after a decade which began with Rachel Carson's (1962) study
of damage to biota and introduced words like "ecosystem" and
"biosphere" into the language. That philosophy holds that
ecological processes are complex and intricately inter-
related--certainly too complex to be relegated to "critter
counts" and reductionist, unelaborated mentions of checklist
physiographic indicators. Thus, the documents written
pursuant to NEPA section 102(2)(C) were not intended to give
short shrift to serious biological impacts.

The contents of EISs stand as strong evidence of the
impracticality of rationalistic reforms of decisionmaking.
Since the 1950s, administrative theorists have generally
believed that uncertainty in predicting the future, the
organizational costs of comprehensive analysis, and the
formadible problems in obtaining agreement on policy
objectives at the beginning of a process make rational-
comprehensive decisionmaking impractical and inefficient.
Spurning this conventional wisdom and embracing the
economists' and engineers' affection for the comprehensive
model, NEPA asked agency staffers to act as if they were
rational decisionmakers--to comprehensively consider
alternatives and their consequences. The contents of the 29
sample EISs confirm our earlier judgment that EISs fall far
short of the rational-scientific ideal (Friesema and Culhane
1976; Culhane, Armentano, and Friesema 1981).

The rationalist ideal of the prescriptive literature,
however, is not unambiguously desirable. Many verbal
forecasts are hopelessly muddled; other clear, thorough,
quantified predictions are excellent models of environmental
assessment. But quantification is not a necessary and
sufficient condition of good environmental assessment.

The air quality section in one of the cases in the
content analysis pretest, for example, fulfills every
technical characteristic of Canter's (1977) model of high-
quality environmental assessment. The Clovis Heights FEIS
examined two subdivisions to be built in a suburb of Fresno,
California. Its air quality section began with a transportation assessment that estimated the subdivisions would
generate additional traffic of 10,920 vehicles per day.
Using handbook figures on emissions per vehicle, these extra
trips were predicted to create additional emissions of
108,560 grams/day of carbon monoxide (CO), 18,218 g/day of
hydrocarbons, 21,367 g/day of nitrogen oxides (NO_x), and
1,052 g/day of particulates. The EIS writers then estimated
emissions from heating furnaces in a similar fashion. Then,
using standard air pollution models, the EIS writers estimated the changes in ambient air quality caused by the

113

subdivisions' emissions, compared forecast ambient levels with air quality standards, and commented on the relative significance of each change. For example, one-hour NO_x would increase from 583 to 652 ug/m^3, compared to a standard of 470 ug/m^3. Up to this point, the section is a model of clear analysis.

The section concluds by proposing mitigations, as required by ideal environmental assessment procedures:

Air quality in the vicinity of the projects will exceed air quality standards under worst-case conditions due to poor ambient air quality, with the projects having only minimal impact on this ambient air quality. Pollutants which require mitigation include oxidants (125% above the federal 1-hour standard), nitrogen oxides (38% above the federal 1-hour standard), and particulates (50% above the state 24-hour standard). Therefore, these pollutants require mitigation.

Recommended Mitigation

1. Refrigerated air conditioning is required for all housing units at each project site.
2. All doors and operable windows must have seals to limit intrusion of untreated outside air. . . .
3. All air intakes to the above air conditioning systems shall be equipped with easily replaceable particulate air filters.
4. Intake of outside make-up air for the air conditioning systems must be held to the minimum allowed by code and good design practice.

Refrigerated air conditioning is required to prevent future residents of the proposed housing from having to open windows and doors on hot summer days to obtain a cooling breeze which would expose them to the full impact of the pollutants in the ambient air (HUD 1977: II.20).

One is at a loss to explain an "analysis" like this. Were the EIS writers serious when they recommended closing windows tightly to keep polluted air outside? Was this passage intended to be an inside joke between the EIS writers and any person persevering enough to read this seven-page section carefully?

The rationalist model of the EIS analysis is not, as noted earlier, the only model of the NEPA process. Friesema and Culhane (1976), Liroff (1976), Andrews (1976), Wichelman

(1976), Fairfax and Andrews (1979), Taylor (1984), and many others have argued that the NEPA process is fundamentally a political-administrative process, not just--if at all--a rational-analytic process. In particular, the NEPA process formally opens up lead agency decisions to public and interagency review. The public review component of the NEPA process was vaguely anticipated by Congress (which only required that EISs be publicly available under the Freedom of Information Act), was clearly intended by CEQ reformers from their earliest EIS guidelines, and is now understood to be part of the EIS game by all players. The writers of the Henrico EIS in the sample noted, for example:

> The primary role of the EIS is that of a public infor-
> mation document. The EIS allows the lead agency . . .
> to not only objectively include all environmental con-
> siderations into the planning process, but to provide an
> opportunity for citizen participation (EPA 1978: I.3).

The NEPA process builds in a tension between the useful-ness of an EIS as the focal point of a public participation process and the expectation that an EIS should be a rational decision document. Since the technician's jargon is formid-able to the layman, analysis that conforms to the strictures of the rationalist ideal of environmental prediction will routinely prove antithetical to public participation in the NEPA process. Recognizing this conflict, federal district court Judge James Burns, in <u>Oregon Environmental Council v. Kunzman</u> (1985), recently held a Forest Service herbicide EIS to be inadequate because its worst-case analysis is "hyper-technical, complex, and replete with lengthy equations and calculations," and thus violated CEQ regulations (1978: §1502.8) that EISs "shall be written in plain language."

Public participation and objective discussion of environmental impacts are logically compatible, but the political features of the NEPA process often induce EIS writers to stray from the path of righteous assessment. Public participation makes EISs into adversarial documents, and the job of a competent adversary is to tenaciously defend his or her client's interests. Reading EISs as adversarial, political documents helps explain some deficiencies of EIS forecasts. Many of the issues of concern to the participants in NEPA reviews lie within the social category: public services and regional planning, which directly affect local governments; land use, social irritants, and residential displacement, which affect a project's immediate neighbors; and amenity values, which EIS

115

writers often seem to think are the primary concern of conservationists. Thus, the imbalance between biological and social forecasts would surprise an advocate of the scientific model of environmental analysis or a student of NEPA case law and legislative history, but coverage of client issues is perfectly apt if EIS writing is driven by anticipation of public and interagency review pressures. Similarly, one must be deeply imbued with a firm sense of pluralist value neutrality to write that a stripmine is something that "other people consider . . . an interesting addition to the landscape." That is, forecasts are imprecise both because agency staffs do not have the time to calculate every impact as specified by the scientific state of the art and because vague forecasts or creative verbiage help paper over unsightly impacts. This explanation, however, should not be confused with approbation.

Finally, these characteristics of forecasts affect an evaluation of EISs' predictive accuracy. This study's original research design envisioned a statistical comparison of the time series of actual postproject observations on some indicator of an impact with the line representing the EIS's forecast about that indicator. If EIS writers are vague or imprecise in their forecasts, they give themselves a wide margin of error. That wide margin poses an interesting conceptual problem—exactly what is the expected postproject trend against which actual events are to be compared? Given the distribution of the sample's 1,105 forecasts on the vagueness index, this problem understandably plays a significant role in evaluating the prescience of the 29 field-sample EISs.

NOTES

1. The Random House Dictionary, unabridged edition (New York, 1966), p. 1133. The two terms have equivalent meanings in certain technical usages, for example, in "econometric forecasting."

2. However, repetitions of the same forecast in different parts of an EIS were not coded as separate forecasts. Significant impacts are regularly synopsized in the summary sheet at the beginning of the EIS, discussed fully in the project impacts chapter, and repeated in the "short-term versus long-term impacts" or "irreversible and

irretrievable impacts" chapters and responses-to-comments appendix. As long as a forecast's characteristics did not change materially from one passage to the next, it was recorded as a single forecast.

3. A second-round coding of the field-sample forecasts (see the methodology appendix) identified one more quantified-probability forecast, but doubling the affected row in Table 4.3 to compensate for this coding error would not change our conclusion about this characteristic of an ideal prediction.

4. The beneficial forecasts' mean index score of 2.08 is significantly less than the adverse forecasts' mean of 2.46 ($t = 4.1$). However, 111 "no impact" forecasts about potentially adverse impacts were coded "beneficial." Deleting "no impact," "neutral," and "ambiguous" impact forecasts plus mitigations and objectives leaves the following distribution:

	Adverse	Beneficial
Quantified forecast	139 (32.6%)	54 (30.3%)
Verbal forecast	288 (67.4%)	124 (69.7%)
Totals	427 (100%)	178 (100%)

Other variables' relationships to the index do not show statistically significant differences in analysis-of-variance tests. Summary analysis-of-variance statistics for these key tests are as follows: category (biological, etc.), $F = 5.73$, $eta^2 = .02$; impacts (mean = 2.45), mitigations (2.22), objectives (2.29), $F = 1.97$, $eta^2 = .004$; quantification (quantified, verbal, no impact), $F = 852.6$, $eta^2 = .61$; and EIS, $F = 3.97$, $eta^2 = .09$.

5. The correlation of the vagueness index with the number of comment letters, deleting the Grand Teton case, is .18, with a broad inverted-U-shaped scatterplot. (It would be senseless to compute a nonlinear correlation of this parabolic N = 28 distribution.)

6. Jackson airport cases include Sierra Club v. Department of Transportation (D.C.Cir., 1982), Sierra Club v Department of Transportation (D.C.Cir., 1985), and Sierra Club v. Department of the Interior (D.Wyo., 1985). None of the decisions in the Sierra Club's series of Jackson airport suits from 1972 through 1985 have been published in the West reporter series or Environmental Law Reporter; summaries can be found in Sierra Club Legal Defense Fund annual reports (e.g., Ruckel 1980, 1985).

7. For readers who do not regard the two "not in my backyard" suits as evidence of major controversy, the

average index scores of the remaining EISs are as follows: Cross-Florida, Jackson, and Monticello, mean = 2.31 (110 forecasts); Cross-Florida, Jackson, Monticello, plus Shippingport, mean = 2.18 (150 forecasts). The litigated projects, Monticello, U.S. 53 - U.S. 8, Cummington, Jackson, and Cross-Florida, average 2.48. In addition to these five projects, in which controversies resulted in extensive or published litigation, some other projects in the sample were subjects of minor litigation such as the I-295 right-of-way suit (see Chapter 3) and some pro forma condemnation suits in the Illinois Beach acquisition.

   8. In addition to the studies mentioned in the text, see the Sewell (1980) and Rosenberg et al. (1981) studies of U.S. and U.S./Canadian environmental assessments.

5
Project Record Keeping and Other Obstacles to Environmental Audits

Would-be environmental auditors face trying methodological trade-offs. Some difficulties encountered in such evaluations stem from the characteristics of EIS forecasts and project impacts. The fundamental trade-off, however, involves a difficult choice between the quality of data that can practically be collected and the diversity of impacts and projects involved in an audit. Once evaluators choose to retrospectively audit a representative range of U.S. EISs, their central problem is that good data on project impacts are quite rare.

This methodological difficulty suggests some substantive conclusions about the comprehensiveness of project management. This chapter serves principally as a methodological preface to the analysis of EISs' predictive accuracy in Chapters 6 and 7, but some readers may be bored by the details of ideal environmental audit methods, fieldwork protocols, and so forth. Such readers may wish to treat this chapter as a midbook appendix, glance at Table 5.1, and skim the description of the study's accuracy classification scheme. They could then concentrate on the final section's discussion of the implications of federal agencies' lack of familiarity with project consequences.

Relatively few environmental postaudits have been conducted to date. As is often true at the beginning of a line of research, the early environmental audits have been case studies. The first and still one of the best environmental audits was the University of Wisconsin study of the Columbia generating station (see Loucks 1982). Most of the studies reported at the 1985 Banff conference on environmental audits were case studies of major Canadian projects' impacts.[1] Two other studies audited four cases apiece:

119

the Aberdeen University evaluation of four British environmental assessments (Bisset 1984) and Hunsaker and Lee's (1985) analysis of four EISs' "worst case" forecasts.

Of these evaluations, the Columbia station case study best illustrates the trade-offs involved in environmental audits. The Wisconsin researchers used a comprehensive theoretical framework that combined mass balance and integrated ecosystem approaches to environmental impacts. (The former maps the fate of all materials entering and leaving the powerplant system, and the latter documents the effects of plant residuals elsewhere in the watershed or food chain.) University scientists began to collect baseline information in 1971, and collected impact data from the 1975 completion of the station to 1981. Data were gathered using monitoring protocols designed to exacting scientific standards. The research team included, at peak, 100 people working in 24 research groups. In short, the Wisconsin team devoted considerable resources to collecting comprehensive, sophisticated data on the plant's impacts. However, like all case studies, the Columbia study offered no assurance that its findings would have been the same if a representative sample of projects had been studied.

This study's research strategy makes different choices from those made by the Columbia study's research team. On the one hand, we chose to audit a moderate size sample of 29 EISs that contains a range of projects and impacts representative of all EISs. On the other hand, we did not wish to spend a decade collecting perfect monitoring data. In addition, no sponsor could afford the cost of a Columbia study design across 29 cases—say, 24 scientists per case, times 29 cases, times 10 years, for 6,960 scientist-years, times a $42,000 annual personnel factor, for $292 million, plus expenses. Instead, we assembled a small staff that, because it evaluated all kinds of impacts, frequently confronted data and phenomena that were far afield from the disciplines of its members' formal training. Most important, we relied on existing sources of data. We were thus at the mercy of the information available in the files of lead agencies, project sponsors, and other parties.

THE IDEAL IMPACT AUDIT MODEL

The basic premise of an environmental assessment is that a project will independently cause some environmental change. A technically trained analyst would operationally define that change as a postproject increase or decrease in some measurable indicator of an environmental condition. The fundamental problem in measuring postproject environmental impacts is the same problem that confronts evaluation researchers generally: How may one validly conclude that a discrete event—whether a natural resources project, a new law, or an education innovation—is actually the cause of an observed change in a variable? Scientists are most confident of the results of experimental studies, but for both practical and ethical reasons most policy-relevant events cannot be subjected to true experiments. Members of a polity cannot be randomly assigned to groups bound or not bound by a new law, for example, nor can the complex environment of a project be experimentally controlled.

Evaluation researchers generally regard the interrupted time-series design as a very sophisticated model for evaluating the many kinds of public policy effects that cannot be subjected to experimental controls. The design does not satisfy all the criteria for a perfectly valid design, but it has fewer weaknesses than any design that does not require random assignment of subjects among treatment groups. Thus, methodologists call it a "quasi-experimental" research design (Campbell and Stanley 1963, Cook and Campbell 1979). The design requires a series of observations of some subject at regular intervals, with the treatment being the interruption at midseries that is associated with any change in the pattern of observations. The change may involve a discontinuous increase or decrease, a change in the slope of the preproject trend line, or some nonlinear mathematical form of the trend.

Since project completion seems to be a discrete event, and natural environments are open systems that are subject to many influences and not experimentally controllable, an interrupted time-series design seems to be appropriate for auditing EIS predictions. The proposal for this study conceived of EIS postaudits as essentially the kind of design depicted in Figure 5.1 (Culhane and Friesema 1982: 6-15). A properly specified prediction should consist of enough predicted values to determine a postproject trend, such as the monotonic increase "P P P P" in Figure 5.1. (The prediction could just as easily call for a steplevel, cyclical, or other form of postproject trend.) An environmental audit, then, simply consists of a comparison of the prediction line with the actual postproject data on the impact. In the example in Figure 5.1, while the set of observations "oooo" shows a postproject increase in the variance of the series, one could not conclude that the project caused a permanent change in the indicator or that

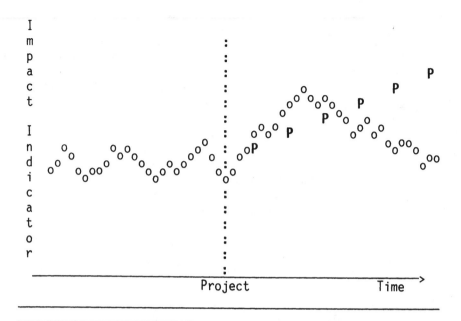

Figure 5.1
The interrupted time-series model of postproject impacts.
Points **P P P P** denote predicted values of an environmental
indicator; points oooo represent actual preproject and
postproject values of the indicator.

the impact was consistent with the prediction.

Environmental postauditing can be a logical extension of
the science-driven version of the rational environmental
assessment model reviewed in Chapter 1. Beanlands and
Duinker (1983) make this connection most clear in their
influential framework for environmental assessment in
Canada. They depict the interrupted time-series model as an
"operational paradigm" of project impacts. Moreover, as
seen especially in the bottom panel of Figure 5.2, they
regard impact prediction and impact audits as integral
components of a rational-scientific approach to environmental assessment. It is an approach conceptualized in
terms of "hypotheses" and "experiments" designed as much to
advance scientific understanding of ecosystem behavior as to
provide information to decisionmakers.

The interrupted time-series design is a strategy for
gathering valid data; it is not a statistical method <u>per se.</u>
Only under very limited conditions can an interrupted time-

a) SIMPLISTIC VIEW OF MAJOR EIA STEPS

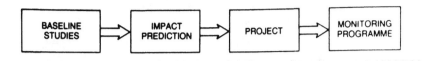

b) AN OPERATIONAL PARADIGM

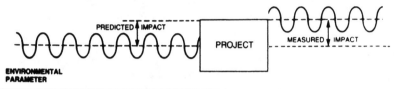

c) THE PROJECT AS AN EXPERIMENT

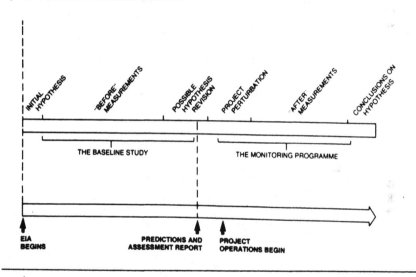

Figure 5.2
The relationship of the interrupted time-series model to general frameworks of environmental assessment: The Beanlands-Duinker "Impact Assessment Paradigm." Source: Beanlands and Duinker (1983: 86), reproduced with permission.

series model be tested using ordinary least-squares regression. The statistical method preferred by most time-series evaluation researchers is autoregressive, integrated moving averages (ARIMA) regression (cf. McCleary and Hay 1980). ARIMA provides a researcher with robust techniques for handling common problems in time-series statistical analysis. However, it requires more than 40 observations to compute its statistical parameters, which poses major barriers to evaluating many impacts whose data are kept on a strictly annual basis. This study's original proposal thus envisioned correlating predicted and actual impacts using a pooled cross-sectional time-series statistical technique. This method generates a large enough sample for generalized least-squares regression analysis by pooling the short time-series of a number of cross-sectional cases (Dean and Stimson 1980, Stimson 1985). Suppose property tax forecasts had been selected for analysis in the audit, and the nature of project implementation dates allowed a minimum of five years of postproject data. Five preproject and five post-project data years would be inadequate for statistical analysis, but 290 observations (ten years multiplied by 29 projects) would be satisfactory. Thus, this research design required, first, properly specified forecasts in each sample EIS about the same impact type and, second, ten years' worth of time-series data in the same units of measurement on the impact indicator across all 29 projects.

FIELD DATA-GATHERING

Even before fieldwork began, the demands of the pooled cross-sectional time-series design proved difficult to satisfy. During a pretest of the content analysis protocol, we learned that it would be difficult to find salient forecasts about the same types of impacts across all EISs in a representative sample. The matrix of 188 impact types and forecast salience in the field-sample EISs confirm this judgment. For example, most EISs contain a forecast about historic or archeological impacts, but this is often an obvious "no impact" forecast because there are no registered historic or archeological sites within the project area. On the other hand, many interesting and salient impact forecasts appear in only a minority of the sample EISs.

In light of the variations in impact types and advice about the probable difficulties in obtaining valid time-series data about impacts on the natural environment, we adopted a more flexible fieldwork protocol than originally

124

proposed.[2]   As a first step, the field sample was selected
to enhance the possibility of finding comparable forecasts
within subsets of the field-sample EISs (see Chapter 3).
That is, it might not be possible to find in every EIS a
forecast about ambient concentrations of criteria air
pollutants (i.e., nitrogen oxides and hydrocarbon emissions), but highway EISs should contain such forecasts.

Data were then sought during fieldwork on eight categories of postproject outcomes: project objectives, mitigation measures, and physiographic, biological, economic,
social, controversial, and unanticipated impacts. Field
teams arrived at each project site with a list of candidate
forecasts in the project's EIS. The list usually contained
two or more forecasts within each of the first seven categories, insofar as the EIS contained salient forecasts in
each category. Controversial impacts proved difficult to
audit, since the coders could detect no significant controversy in eleven of the EISs in the sample. Occasional EISs
also contained no sensible forecast within some other
category. For example, the EIS on the Grand Teton National
Park master plan did not contain a biological forecast,
which is quite surprising in light of recent controversies
(Chase 1986) over park management and wildlife habitat in
northwest Wyoming, so the field team had no choice but to
skip this category during the Grand Teton fieldwork.

Field teams conducted fieldwork during 1983. The teams
sought data from existing sources, including the files of
the lead agency, local government, regulatory agencies,
project sponsors, and chambers of commerce. The teams
sought time-series data from 1970 or earlier through 1982,
but acquired a variety of non-time-series data as well.
While in the field, the research teams also conducted 120
interviews with informants who were familiar with the
projects. (The fieldwork procedures are described more
fully in the methodological appendix.) The interviews
covered informants' observations about impacts generally,
although the primary purpose of the interviews was to identify any unanticipated impacts.

The 29 lists contained a total of 370 forecast impacts.
(The audit's units of analysis, forecast impacts, are
defined at the end of the chapter.) Fieldworkers did not
seek data on all these cases; if satisfactory data were
obtained about one or two economic impacts of a project, for
example, fieldworkers might not seek data on low-priority
economic forecasts on the EIS's list. Thus, data were
actually sought on 339 impacts during fieldwork. The more
salient forecast impacts were almost always sought first in

Table 5.1
Summary of accessibility of data during fieldwork. Table lists the primary types of data for forecast impacts; secondary data types are shown in Table 5.3. Cases listed in rows 1-5 have data rated "exact," "adequate," or "low adequate"; any data represented by entries in rows 6 and 7 are rated "inadequate." N = 29 projects.

Accessibility	Category of Forecast Impact				Totals
	Phys.	Biol.	Econ.	Social	
Data/Information Acquired					
Time-series or time-ordered indicators	20	3	42	24	89
Other quantified data	5	6	9	17	37
Nominal status information	17	9	1	17	44
Verbal or documentary information	10	15	13	31	69
Subtotal, data or information acquired	52	33	65	89	239
No Data or Only Inadequate or Flawed Data	22	8	12	6	48
Only Inadequate Interview Opinions	22	7	6	17	52
Subtotal, impacts sought in the field	96	48	83	112	339
Forecasts on Candidate Lists, But Data Not Sought During Fieldwork	7	2	4	18	31
Total, all candidate impacts in fieldwork	103	50	87	130	370

the field, so the 31 cases not sought dealt with relatively uninteresting forecasts.

DATA AVAILABILITY

The information available about the impacts of the 29 field-sample projects is quite varied. As shown in Table 5.1, fieldworkers failed to obtain satisfactory data on 100 impacts, or 30% of the impacts for which data were sought during the fieldwork. In half of these cases field crews could locate either no data at all or only fatally flawed data. The only information on another 52 impacts consists of unsubstantiated interview information, often no more than an interviewee's assurance that there had been "no problem" regarding the impact in question.

In other words, no one--not lead agency officers, not project sponsors, not some other government officials, and certainly not the authors--possesses very solid information about almost a third of the impacts subject to our audit. Hereafter, we shall refer to the 239 cases for which more or less adequate data were obtained as the "field sample." The "field sample," however, really consists of 339 impacts. Methodologically, our findings about the relative accuracy of EIS forecasts must be qualified because we cannot classify the missing 100 cases. These forecasts are probably about as accurate as the 239, but they might not be. More important, in the real world of resource management, the officers who are supposed to be responsible for these projects are relatively ignorant about them.

Field Data Characteristics

Numeric data of the form demanded by the ideal impact audit model are difficult to obtain. In only 81 cases, or less than a quarter of the full field sample of 339 forecast impacts, were minimally satisfactory time-series data available. Systematic time-ordered data were available for another eight cases and were usually quite good. (However, as discussed more fully below, they do not match the form of the interrupted time-series model.) If non-time-series data were available, the fieldworkers took them. As indicated in Table 5.1, 37 impacts were audited based on quantified but non-time-series data. Thirteen cases involve some form of preproject, postproject quantified observations. Another five data sets consist of the results of controlled studies,

127

Table 5.2
Impact types of the 239 field-sample cases. The field
sample contains 102 distinct impact types, though several
types were combined to simplify the table.

Impact Types	N	Impact Types	N
Physiographic Impacts		**Biological Impacts**	
Air Quality		Wildlife Habitat, Populations,	6
Criteria pollutants	5	or Misc. Wildlife Impacts	
(CO, SO_x, NO_x, etc.)		Aquatics	
Fugitive dust	1	Fish habitat, general aquatics	4
Fluorides	1	Fish, entrainment, entrapment	3
Water Quality		Benthic populations, habitat	4
Turbidity	3	Aquatic microorganisms	1
BOD, sewage	3	Birds	
Thermal effects	3	Habitat	2
Toxics	1	Avian diseases, public	1
Limnology, eutrophication	1	health effects of	
Groundwater quality	2	Forests, Trees, and	6
Other, multiple pollutants	2	Silvicultural Impacts	
Hydrology and Soils		Vegetation and Grass	3
Erosion	6	Wetlands	2
Surface water flow	2	Herbicide Exposure, Fauna	1
Flooding	2		
Soil quality (metals, pathogens)	1		
Channel alignment	1	**Social Impacts**	
Reservoir sedimentation	1		
Soil conservation	1	Population	6
Radiation		Demography	1
Ambient (air or water)	10	Government Services	
Rad. waste storage/safety	2	Water supply (municipal of M&I)	3
Fire hazard/management	3	School enrollment	2
Solid waste disposal, flyash	1	Police	1
		Public transit use	1
		Social service, senior citizens	1
		Sewage load	1
		Government service, general	1

Economic Impacts

Employment, Direct or Secondary Sectors	9
Agricultural production	6
Electricity generation	4
Timber production/sales	3
Retail sales	2
Homebuilding industry	1
Grazing use	1
Natural gas production	1
Government Finance	
Property tax revenue/base	11
Sales, other local taxes	3
Project cost	4
Federal payments in-lieu of property taxes	2
Business Development	
Commercial development	2
Industrial development	1
Recreation business development	2
Economic development, general	3
Cargo Shipping, Waterborne	4
Air Passenger Volume	1
Research and Development Engineering	1
Product Market Diversity	1
Property Values	1
Access to Businesses, Disruption of	1

Social Impacts (continued)

Social Welfare	
Noise	6
Blasting vibration, safety	2
Public safety, nonoccupational	2
Public health	1
Amenities	
Aesthetics	6
Hunting or fishing	3
Yachting	1
Wilderness	1
Open space preservation	1
Recreation visits, use, facilities, miscelaneous	9
Traffic	
Auto volume, traffic flow	10
Safety, auto	5
Air traffic safety	1
Construction period, access or traffic flow	2
Parking	1
Regional transportation plan	1
Housing	
Displacement, residential or commercial	7
Neighborhood cohesion, privacy	2
Housing quality, maintenance	1
Residential development	1
Land Use	
Land use, land use conversion[a]	2
Land taking for project	2
Archeological, Historical Sites	4
Scientific Research	1
Legal Compliance with Government Regulation	1
Total	239

[a] One land use case classified within the economic category.

and these are excellent sources of information.

The remainder of the 239 cases were evaluated primarily on the basis of nonquantified information. Almost a fifth of the 239 cases involve distinct statuses measured by dichotomous or nominal measures. For example, the outcome of a mitigation can be aptly indicated by whether the mitigation was done or not. A dichotomy is a technically a binary variable, but most readers will recognize these cases as simply nonquantitative information. A few cases could be audited on documentary evidence, such as aerial photographs and inspection reports. However, the primary source of data on 64 cases consists of verbal information, that is, the statements of interviewees.

The field data also cover a wide range of impact types. Of the 188 impact types encountered during the content analysis, 102 are represented among the 239 field-data impacts. As shown in Table 5.2, most impact types are represented by only one to four cases. Among the natural environmental impacts, only radiation accounts for a substantial number of cases. The half-dozen erosion cases mostly involve routine mitigation measures. The half-dozen wildlife cases are divided evenly between habitat and population impacts (which are distinct coding categories, even if the impacts differ little in practice). The socioeconomic impacts are even more varied. Three government finance impacts (property taxes, other local taxes, and federal payments in lieu of taxes) constitute the largest set of similar impacts, involving the same basic economic consequences and unit of measurement. Social impacts represent 39 impact types--the most diverse category in the sample.

The units of measurement of the field data are as diverse as the range of measurement units--a total of 53 measurement units. (See Table 5.3.) The most common measure is dollars, the key economic unit. Otherwise, the measures closely parallel the field sample's impact types: 15 data sets measured in parts per million or grams per cubic meter/liter, six measured in curies, six wildlife population counts (or, in the trade jargon, "critter counts"), and so forth. The most common measurement unit is "none," since 70 cases' primary information is verbal.

Table 5.3
Primary units of measurement of field data. The table lists the units of measurement of the primary types of data; 102 cases' data sets also involve more than one distinct measurement unit (not shown). As in Table 5.2, a few similar measures (e.g., ppm, mg/l) are combined.

Unit of Measurement	N	Unit of Measurement	N
Physiographic Measures		Social Measures	
PPM, mg/l, or similar	12	Average daily traffic (ADTs)	9
Curies, rads, or similar	7	Buildings, residential units,	7
Water flow measures (e.g., cfs)	3	building permits	
Turbidity units (e.g., JTUs)	2	Traffic accidents	6
Temperature (re thermal impacts)	2	Average daily traffic (ADT)	5
Fires, number of	1	School enrollment	5
Coliform count	1	Population	3
Fissile inventory ratio	1	Acres (social impact)	2
		Inch/sec., ground movement	2
Biological Measures		velocity	
Population count, any species	3	Decibels	1
Species, numbers of	2	Fishing licenses	1
Species index (e.g.,	2	Game kills	1
Shannon index)		Law enforcement officers	1
Acres (biological impact)	1	Gallons per day	1
		Recreation facilities, number	1
Economic Measures		Housing management rating scale	1
Dollars	26	Wilderness, number	1
Employees	8		
Megawatts	3	Other	
Ships, watercraft	3	Percentages, misc.[a]	3
Board-feet (MBF) of timber	3	Mitigation—amelioration	21
Livestock, AUMs or numbers	2	Mitigation—monitoring, studies	10
Firms, number of	2		
Tonnage (i.e., cargo)	2	No measurement unit, just	
Passengers	1	restatement of impact type	70
		Total	239

[a] Five other cases with data in percentage form are listed elsewhere with their divisors' measurement units.

131

Field Data Validity

The weaknesses inherent in the varied character of the field data is somewhat offset by the fact that the field crews obtained multiple types of data on many impacts. Two types of data were acquired on 101 impacts, and three types on another 22. (See Table A.2 in the appendix.) The largest set of secondary data consists of interview information. In almost all cases these interviews supplement primary data that are quantified or nominal. The secondary data also include 37 sets of quantified data, half of which serve as a numeric backup to verbal primary information.

In addition, we regularly obtained multiple and comparison indicators about a given impact. (Multiple indicators are not reflected in Table 5.1 or A.2. The entries in the tables show the numbers of forecast impacts whose data are of a certain type, not the numbers of indicators.) For example, the data set for a single forecast impact about a property tax base might consist of six time series on the assessed valuation and tax receipts of the project's township plus valuation and receipts for the county and a nonimpacted township, or the data about a water-quality impact might consist of ambient biological oxygen demand and phosphate and nitrate concentrations at several monitoring sites upstream and downstream from a sewage treatment plant.

The fact that some verbal information is backed up by secondary quantified data indicates that the validity of these varied data sets entails more than the raw distinction between quantified data and nonquantified information. Several factors affect the utility of the field data for assessing the match between forecast and actual impacts. These include the nature of the primary data, the amount of secondary data about an impact, the intrinsic validity of a a quantified indicator, flaws in a set of numeric data, the correspondence between units of measurements and populations of the forecast and data, and the reliability of interview information. These factors are summarized in a three-code data aptness scale (detailed in the methodology appendix): "exact," "adequate," and "low adequate." Data deemed fatally flawed or unreliable were rated "inadequate" and excluded from further analysis; see rows 6 and 7 of Table 5.1. As shown in Table 5.4, about four fifths of the data sets were rated either "adequate" or "exact."

Nonquantitative data proved perfectly appropriate for auditing the accuracy of several forecasts. For example, the interview information on one "neighborhood cohesion"

Table 5.4
Quality of field data by selected case characteristics. The "exact" rating includes 6 data sets rated "better."

Case Characteristic	Data Aptness			Total
	Exact	Adequate	Low Adeq.	
Primary Data Type				
Time series, time-ordered	54	16	19	89
Experimental results	3	1	0	4
Preproject, postproject quantified	6	6	6	14
Other quantified	6	4	9	19
Nominal status	15	22	7	44
Documentary evidence	4	1	0	5
Verbal, interview	9	44	11	64
Impact Category				
Physiographic	23	22	7	52
Biological	9	18	6	33
Economic	29	22	14	65
Social	33	32	21	89
Forecast Type				
Impacts	68	60	32	160
Objectives	17	17	10	44
Mitigations	12	17	6	35
Project Type				
Energy	26	16	7	49
Housing/urban/buildings	23	28	7	58
Water resources	19	19	11	49
Public lands	16	15	7	38
Highways	13	16	16	45
Totals	97	94	48	239

impact is rated "exact" largely because of reliable interviews on what is essentially a subjective phenomenon. Four cases with documentary evidence also received "exact" data aptness ratings; this evidence includes aerial photographs of forest blowdown impacts and extensive newspaper coverage of a firing range incident. Interview and documentary information is also sufficient to establish the status of most "nominal" impacts, such as whether or not a mitigation measure was implemented; nominal data were generally rated only "adequate" unless the information was supplemented by good secondary data or documentary evidence. On the other hand, some quantitative data were rated "low adequate." A dozen time series were rated "low adequate" because of mismatches between the populations from which the data were drawn and those of the relevant forecasts. Some non-time-series quantified data proved perfectly valid. For example, auditing a project cost forecast requires only a single datum. But more often the data in the "other quantified" row were of low quality.

In general, however, quantified and especially time-series data seemed more valid than interview information for auditing forecasts. Three fifths of the time-series cases received the best rating, "exact," while most cases relying on interview information were rated only "adequate."

Biases in the Field Data

The field data, in addition to being of varied quality, contain some systematic biases. The most important differences in data quality involve the four categories of impacts, and these in turn are often related to types of projects and lead agencies involved.

Socioeconomic data proved relatively easy to acquire. Indeed, in choosing to rely on a strategy of seeking extant data, we were influenced by experience gained in a previous study of the socioeconomic impacts of natural disasters (Friesema et al. 1979) that had successfully sought time-series data on an array of social and economic indicators. Some form of useful information was obtained on 79% of the social and economic impacts audited during fieldwork. Economic data proved to be both relatively accessible and of generally high quality. The economic category has the highest percentage of data rated "exact," and accounts for almost half the time-series data sets in the field sample. Wealth is valued sufficiently, it appears, for people to keep good records on the subject. Interested parties like project sponsors and lead agencies are less useful sources of good economic data than local governments and chambers of commerce. Twelve of the 40 economic time series, for example, are tax data series, chiefly annual property tax data obtained from local assessors' records.

The field teams acquired more information on social impacts than on any other category of forecast. In fact, fieldworkers often filled their quota of social cases so easily that they could afford to ignore 18 low-priority candidate social impacts. The social data they acquired, however, are not quite so robust as the economic data in the sample. Almost half the cases audited based on verbal or documentary primary information fall into the social category and a lower proportion of social data sets were rated "exact" than were their economic counterparts. The large number of social impacts in the field sample is chiefly attributable to the prevalence of social forecasts in EISs; 40% of the 1,105 forecasts in the sample EISs dealt with social impacts, while 35% of the candidate impacts and 37% of the 239 data sets fell into the social category.

Good data on natural environmental impacts proved much harder to locate. Satisfactory physiographic data could be located on only 52 of the 96 candidate impacts sought in the field, a .542 batting average that would be grand in the major leagues but is anemic by normal field research standards. The physiographic data, however, have the virtue of generally being either good or bad: with 44% of the 52 data sets rated "exact" and only 13% rated "low adequate," the physiographic data are the strongest of the four categories in the field sample.

The biological data, with only 33 cases--barely one impact per EIS in the sample--proved even more difficult to locate. First, as noted in Chapter 4, the sample EISs contain few good biological forecasts. Biological impacts represent only 13.5% of the candidates sought during field-work, versus 15% of all forecasts in the sample EISs. Essentially, too many biological forecasts in the sample EISs are unevaluable, which exacerbates the problem caused by the low number of biological forecasts. Second, only nine biological primary data sets are quantified, and only nine (not all the same cases) are rated "exact." The four "exact" quantified biological data sets are so rare they can be described individually: a master's thesis on vegetation effects of land treatment of sewage effluent, a professionally executed study of benthic effects using the Shannon species diversity index, a good study of aquatic habitat in a stream before and after highway construction, and a multi-

135

year time-series monitoring of aquatic microorganisms' species numbers and Shannon indices. Third, given the paucity of quantified biological data and the intrinsic limitations of interview information about biological systems, the 33 biological data sets possess one of the weakest data adequacy distributions in Table 5.4.

The weaknesses of the field sample's natural environment data are further complicated by the fact that most of the good natural science data come from the four nuclear energy projects in the sample. Ten of the 20 physiographic time series deal with nuclear facilities' impacts. The nuclear subsample also produced three of the six best biological data sets. Nuclear projects are closely regulated by the NRC, which requires plant managers to produce extensive, systematic, technically competent, and regular data on many types of impacts. The Alma station had also been involved in several controversies with the Wisconsin Department of Natural Resources, which monitored the plant fairly closely, and this project contributed another six "exact" data sets. Thus, data proved to be relatively abundant and of very high quality on the energy cases.

Two other classes of projects contribute to the sample's imbalance between natural environmental and socioeconomic data. Socioeconomic impacts tend to be more salient than biological and physical impacts in both the highway and urban development subsamples of cases. The housing-urban-buildings data sets are generally good, while the highway cases have the worst data quality ratings of any subsample listed in Table 5.4. Both sets are also heavily laden with social impacts. The six highways and four urban development projects account for 64 socioeconomic data sets (including half the social impacts) in the field sample, versus only 18 physiographic or biological cases.[3] Moreover, half of the highways' physical and biological cases involve mitigation measures; that is, field crews, unable to locate good impact data, could only check on such mundane mitigations as whether construction crews had used hay bales to minimize erosion and siltation from scarified right-of-way land.

The water resources and public lands projects' data are about average. The data on a few of these projects are as bad as those on the typical highway project, while some others are quite good. In particular, the Paint Creek dam yielded acceptable data on 14 impacts, the highest number in the field sample. The Paint Creek data are generally good, including seven time series, three of which deal with phy-siographic impacts. In other words, Paint Creek plus the five energy projects account for two thirds of the time-

136

series data on natural environmental impacts.

MEASURING FORECAST ACCURACY

Several aspects of environmental assessments preclude the use of a pooled cross-sectional time-series measurement model. This state-of-the-art method for evaluating environmental impacts makes certain assumptions: that the forecast will be in the form of a time series, as in Figure 5.1; that fieldworkers could acquire time-series data for five years before and five years after project implementation; and that, within a given category of cases (e.g., physiographic impacts), the forecasts and data would deal with the same impact type and unit of measurement across all 29 projects. In the real world of policy research, evaluators must occasionally compromise a methodological fine point. However, environmental auditors' data are so crude that they must improvise as they go along.

Barriers to Pooled Statistical Audits

The pooled cross-sectional interrupted time-series model of project audits proved inappropriate for six reasons. First, EIS forecasts are rarely specified in the form of a time series. (See Table 5.5.) Our coding of these forecasts was most liberal: we preferred the forecast to have two baseline numbers so we could determine what preproject trend the forecaster believed was occurring, but we required only a minimum of two postproject predicted values for the passage to qualify as a "time-series forecast." There are many methodological pitfalls inherent in working with a forecast composed of two posttreatment data points. (See Campbell and Stanley 1963 and Cook and Campbell 1979.) At a minimum, such a criterion assumes that all impacts will be linear, and there are few good reasons to expect environmental impacts to be linear over time. In any case, only 1.9% of the forecasts in the field-sample EISs are in time-series form; in fact, less than a quarter of the forecasts are quantified at all. The field teams obtained data on half the time-series forecasts in the sample EISs, but these ten cases represent only 3.5% of the field-sample impacts.

Second, EIS writers do not make forecasts about the same types of impacts across all EISs, or even across all EISs on the same general class of projects. Thus, as shown in

Table 5.5
Quantification of forecasts audited in field sample. Secondary forecasts bear on essentially the same impact, but some characteristic of the forecast differs from the primary forecast. Percents in column 5 are taken from Table 4.3.

Forecast Quantification	Primary Forecast	Secondary Forecast	Totals, Field Sample N	Totals, Field Sample %	Percents, All 1,105 Forecasts
Quantified: Time series	9	1	10	3.5%	1.9%
Single-number forecast	46	6	52	18.4%	14.8%
Multiple indicator forecast	20	1	21	7.4%	4.6%
Range-of-values forecast	13	2	15	5.3%	1.5%
Percentages	3	0	3	1.1%	0.8%
"No impact" forecast	15	4	19	6.7%	11.1%
Verbal, unquantified forecast	132	32	164	58.2%	65.2%
No relevant forecast[a]	1	0	1	0.3%	n/a
Totals	239	46	282	100%[b]	100%

[a] That is, an unanticipated impact.
[b] Does not total to 100% because of round-off error.

Table 5.2, there are no impacts common to more than a half dozen or so of the 29 field-sample projects, despite our conscious effort in drawing the sample to select cases that would maximize the chance of finding common impacts.

Third, time-series data exist for relatively few impacts. Indeed, adequate quantitative data on project impacts—time-series or not—are difficult to locate.

Fourth, the time-series data that can be located are distributed in a very skewed fashion among the cases in the study's sample. Most of the good time-series data on natural environment impacts deal with the four nuclear plants plus one dam, only a fifth of the sample projects. It would be invalid to statistically audit forecast accuracy on the basis of a time-series database with these biases.

Fifth, the interrupted time-series model of environmental assessment is misleading or inadequate in many cases. Many forecasts deal with fundamental status conditions, "nominal variables" in statistical jargon. Whether or not a mitigation measure was carried out could be translated into a binary variable graphed on an interrupted time-series chart; as former President Nixon might put it, we could do that, but it would be wrong. Moreover, some perfectly fine quantitative data could not be made to fit into a preproject, postproject interrupted time-series impact model. Such cases include construction period impacts, where the project completion date is not an "interruption" at the middle of a relevant time series; rather, the beginning and ending of construction form the bounding dates of a short time-ordered data series. As another example, a project's total cost is not even a variable, simply a single number.

Sixth, many verbal forecasts deal with complicated processes that could not be easily or adequately represented by a simple one-variable time-series graph. The field crews sought to audit quantified forecasts, of course, and the field sample's forecasts are more quantified than the 29 EISs' 1,105 forecasts as a whole. Nonetheless, most of the field-sample forecasts are still verbal. Reasonable operational indicators could be created for some of these forecasts, but others can be most parsimoniously evaluated by comparing verbal information with a verbal forecast.

A Forecast Accuracy Classification System

In place of the statistical measurement model, we devised a classification scheme to summarize the match between an EIS's forecast and the actual impact discerned during fieldwork. The environmental assessment community has yet to refine any frameworks for auditing the accuracy of environmental forecasts. There are many frameworks for environmental assessment. But as suggested above, it is a far fall from the clean theoretical frameworks into the muck of real-world environmental impacts. The two best previously published environmental audits, by the University of Wisconsin (Loucks 1982) and the University of Aberdeen, [4] essentially relied on impact-by-impact conclusions.

Each classification in the system described below represents the conclusion of a case-by-case evaluation of the data available on a given impact. Quantitative data were evaluated using what Donald Campbell [5] calls the "interoccular trauma test"--that is, does the impact hit you between the eyes? Mitigations and other "nominal" forecasts were evaluated based on interview or documentary evidence about the status of the mitigation or impact, that is, whether it happened or not. Verbal forecasts about complex

139

or subjective impacts were evaluated using as reliable interview or documentary information as could be located.

The correspondence between EIS forecasts and actual impacts cannot be sensibly summarized by straightforward conclusions such as "right" or "wrong." One encounters too many cases incompatible with dichotomous choices--too many "not exactlys." Thus, the classification scheme is inelegant, consisting of 40 classification codes distributed across three distinct dimensions of predictive accuracy.

The first classification deals with the <u>direction</u> of the impact relative to the forecast direction. The two basic direction codes are "correct," which indicates that the impact is in the same direction as predicted, and "wrong," which indicates that the actual postproject trend moves in the opposite direction from that predicted. Three codes indicate conditions in which it is not clear that an indicator has actually changed direction: "continuous preproject, postproject trend," indicating a constant-slope time series that appears unperturbed by the project; "no real change," indicating a stable time-series or other quantified indicator; and "excessive variance," indicating a data series with regular fluctuations significantly greater than any postproject change in the indicator's mean value. Finally, the system contains two codes to cover a number of cases whose impact is nondirectional. For example, the impact "project cost" consists of a single postproject datum that has no meaningful preproject "baseline" number; estimated cost is nothing more than a forecast. Thus, if a project costs more than predicted, one cannot say that the impact is in a correct or incorrect direction. Instead, we encode the direction of such cases "good" or "bad."

The key classification of a forecast's accuracy is called the <u>match</u> between the forecast and the actual impact. The variety of match classifications stems in part from a self-imposed criterion that we bore the burden of proof to demonstrate a forecast wrong. Thus, the scheme includes several codes describing different types of "grey area" forecast inaccuracy.[5] The system contains 19 match codes:

1. "Close"--This classification, the most accurate, indicates that either quantified or verbal data show the actual impact to be the same as that forecast; a quantified impact need not be exactly the same as forecast, just so close that the difference is neither materially nor statistically significant.
2. "Inconsistent"--Impacts are basically opposite from the predicted outcome, with none of the qualifiers

below; quantitative indicators of such outcomes should be in the "wrong" direction.

3. "Exceeds" and
4. "Less"--An impact's indicator is numerically higher or lower than the value predicted, but is in the general direction forecast.
5. "No clear impact, no impact forecast" and
6. "No clear impact, some change forecast"--In cases of "no impact" forecasts and forecasts of some change, no project-related impact is clearly discernible, usually because of data limitations (e.g., an "excessive variance" data series).
7. "Not yet"--A predicted outcome had not occurred by the time of the fieldwork, but field information suggests the outcome is still scheduled or possible.
8. "Complex, basically accurate" and
9. "Complex, basically inaccurate"--In these two classifications auditors could not make a clearcut conclusion because of a definitional problem about an impact, but in their informed judgment the forecast is arguably correct or incorrect.
10. "Impact within the range of a vague forecast"--This important classification covers cases where the impact is relatively clear but the forecast is too vague to be rated accurate; that is, the actual impact falls within the range of possible impacts allowed by the vague forecast.
11. "Unanticipated, adverse" and
12. "Unanticipated, beneficial"--EIS writers failed to forecast these actual impacts, with serendipity distinguished from adverse unanticipated impacts because the EIS process is designed to prevent the latter.
13. "Underanticipated, adverse" and
14. "Underanticipated, beneficial"--These codes cover passages that are so evasive that a reasonable coder might not even recognize them as forecasts.
15. "Dispute"--An inherently subjective impact or outcome is disputed by informants.
16. "Accuracy intuitively obvious"--These impacts must logically occur if the proposed project is implemented; such "incidental impact" forecasts were audited only in the absence of good forecasts within a category during fieldwork.
17. "Spurious"--In several cases an apparent impact proved to be caused by something wholly unrelated to the project.

18. "Mitigation done" and
19. "Mitigation not done"--These secondary classification (see below) codes indicate that a mitigation
measure was or was not carried out.

The third component of the accuracy classification
scheme is an assessment of the relative beneficiality of the
actual impact. This set of codes is naturally an extension
of the four basic forecast adverseness-beneficiality codes
shown in Table 4.5. Because of various possible combina-
tions of impacts relative to forecasts, the beneficiality
classification consists of 12 codes. Certain codes are
self-explanatory: "more beneficial" (i.e., than predicted),
"as beneficial," "less beneficial," "more adverse," "as
adverse," "less adverse," and "as neutral." A few cases of
forecast benefits but adverse actual impacts, or vice versa,
were coded "crossover, adverse" or "crossover, benefit."
Two codes cover situations in which reasonable people could
disagree about the intrinsic beneficiality of an impact:
"subjective" indicates a public controversy about the
impact, and "ambiguous" denotes relatively noncontroversial
cases that the investigator could not code without making a
value-laden judgment. Finally, the code "trivial" indicates
cases in which impacts are not so much adverse or beneficial
as inconsequential.

The accuracy classification scheme also allows for both
primary and secondary codes. The primary accuracy codes
represent the main conclusion about a given case. A case
could receive secondary codes under two circumstances. A
secondary code could merely involve an elaboration on the
primary code; for example, a primary match code of "close"
and a secondary match of "spurious" indicates that a
predicted outcome occurred, but for the wrong reason. In
addition, several cases in the sample involve two related
forecasts. A case might include a forecast about erosion
impacts and a closely related erosion-control mitigation
measure, for example, or a pair of forecasts on the short-
term and long-term effects of a project on property taxes.
In such cases the primary accuracy codes apply to the
primary forecast and the secondary accuracy codes to the
secondary forecast. In other words, this audit's cases or
units of analysis are forecast impacts, each of which is the
subject of one or more forecasts and may be evaluated using
several quantitative indicators or interview sources.

This classification scheme is complex, if not tortuous--
primary and secondary codes about forecasts' directional
accuracy, accuracy "match," and relative beneficiality. It

142

sacrifices the statistical elegance of a pooled cross-sectional interrupted time-series model in favor of a taxonomy that reveals all the warts inherent in EIS forecasts, available environmental indicators, and projects' actual impacts. As the product of 239 case-by-case judgments about diverse forecasts and impacts, the scheme can best be understood by following it through a number of example cases. Chapter 6 will be devoted to that exercise.

AGENCY SELF-MONITORING AND RATIONAL ENVIRONMENTAL ASSESSMENT

A dearth of hard data hinders environmental auditors. Statistically inclined environmental evaluators will naturally be drawn to a methodology such as the pooled interrupted time-series design described earlier. Our hopes to use such a design to audit the accuracy of EIS forecasts were dashed. Data-hungry researchers, true to the rule that any citizen believes that subsidizing his or her line of work is synonymous with the public interest, commonly bemoan the lack of really good databases. However, if the only implication of the lack of good data is that a few number-crunchers are disappointed, it has little social significance.

The state of information on actual environmental impacts, however, helps document the arational realities of environmental management and decisionmaking. The best rationalist theories of both systems analysis (Brewer and deLeon 1983) and environmental assessment (Beanlands and Duinker 1983) incorporate monitoring and postdecision evaluation as essential elements of the decisionmaking process. These theorists do this for two reasons. First, as sophisticated technical analysts, they inevitably adhere to the dictates of the scientific method of their disciplinary training. An absolutely fundamental premise of the scientific method is that, to be confidently used, a theory must be validated or refined by being tested against empirical reality. Environmental assessment predictions embody more or less explicit cause-and-effect theories. If the practice of environmental assessment is to correspond to the rational ideal, EISs writers simply must know which of their assessment concepts are valid, which must be refined, and which discarded.

Second, postdecision monitoring serves an even more important management function. Within the rationalist paradigm, decisions supposedly result in the choice of the

143

optimum project, that is, the one with the highest net benefits. Monitoring, within such a paradigm, ensures that each outcome conforms to the level predicted during the predecision analysis, and thus that the project actually achieves its intended net benefits. The concept of post-decision audits actually developed as a business management tool for precisely this reason--to ensure that the inter-related outcomes that contribute to a program's predicted profit targets actually occur in accord with a firm's strategic plan. Auditing thus represents the follow-through in a "management by objectives" strategy (Drucker 1976).

Monitoring and audits serve the same purpose in environmental assessment. Environmental managers tend to think of monitoring as strictly a tool for detecting significant adverse impacts that would then be subjected to mitigation measures. The initial application to environ-mental management of business concepts of audits, in fact, involved firms' audits to ensure that environmental regu-lations governing pollutants and risky materials were adhered to (A. D. Little, Inc. 1981, Greeno 1983, Harrison 1984). Moreover, for a while it appeared that some form of auditing might become the subject of a new theory in environmental litigation (Environmental Law Institute 1980). But audits are logically just as applicable to project objectives and any nontrivial predicted benefits as to adverse impacts.

Our experience indicates that the federal agency mana-gers of the field-sample projects rarely conform to the model of the self-evaluating administrator. The absence of self-evaluation can be seen throughout the classifications detailed in Table 5.1. First, adequate information cannot be located from any source on almost a third of a randomly based sample of project impacts. Moreover, much of the information that is available is unquantified and marginally adequate. Second, lead agency managers are not systemati-cally aware of much of the information on project impacts that does exist. The field teams acquired their data from many sources, not just from sponsors' or lead agencies' project managers. Those managers frequently knew little or nothing about the impact data that did exist except for that in their own files. So the level of managerial self-awareness is much lower than suggested by the 70% data-availability rate shown in rows 1-5 of Table 5.1.

Lead agencies are systematically ignorant about several specific classes of impacts. Highway EISs routinely note that local increases in air pollutants and noise levels will be a significant impact of the project. These documents

144

regularly quantify these forecasts, often based on field monitoring of preproject baseline conditions. However, none of the six field-sample highways' data sets contains a decent time series, preproject/postproject readings, or any other quantified monitoring of project impacts. Taken together with the often haphazard monitoring of traffic volume, which is certainly not peripheral to the highway department's mission, the highway cases richly deserve their place at the bottom of Table 5.5's data quality ratings.

Most important, agency officers usually have no decent indicators of biological impacts and no access to such indicators from any other source. As was also noted in Chapter 4, the NEPA process was intended to protect the natural environment, especially biological processes. Thus, not only are EISs' biological forecasts the weakest of the four categories, but agencies have the weakest information on the actual biological impacts of their projects.

A few gaps in agency recordkeeping are difficult to reconcile with any notion of a properly functioning public bureaucracy. For example, the Forest Service's primary commodity product, especially in the Pacific Northwest, is timber. The service, which is widely regarded as one of the best-administered agencies in the federal bureaucracy (cf. Kaufman 1960), has been immersed for a decade in land use planning processes that demand reams of inventory data. Thus, we could not understand how the Snoqualmie National Forest staff did not possess and could not provide timber sales data for the planning unit of the Weyerhaeuser timber road case for more than the two years prior to our field-work. Nothing, it seems, could have been more central to the mission of the agency, and particularly this forest, or more useful to the service's officials at that time.

The major exception to these generalizations is the class of nuclear projects regulated by the NRC and its predecessor, the AEC. Decision theorists such as Lustick (1980) have argued that rationalist decision practices are most useful when a bad decision would be extremely costly. Thus, for example, decision analysts in the strategic nuclear weapons community act as if they were rational decisionmakers. The civilian nuclear power program seems to operate on a similar logic: the cost of a significant accident involving nuclear plants is deemed so dire that many rationalist policy practices are followed. Certainly these plants are much better monitored and audited than the other 25 projects in the field sample.

As a general rule, agency decisionmakers seem preoccupied with completing specific projects. They process whatever analyses and paperwork, including NEPA's environmental assessments, are necessary to obtain approval for their projects. Then their attention shifts to their next assignment. There are few incentives in a project-by-project system to audit actual project impacts. Impact monitoring is moderately expensive, the results contribute little to approval and construction of the agencies' menu of projects, and audit results might even threaten an agency's program. In the absence of any inducements linked to actual performance, a decisionmaker is likely to be averse to the substantial risk that an audit will prove embarrassing by documenting a project's shortcomings. Thus, with few positive incentives to self-evaluation and substantial risks, agency managers seem to live by the maxim that ignorance is bliss.

NOTES

1. Follow-up Audit of Environmental Assessment Results Conference, Banff Centre, Banff, Alberta, Canada, October 13-15, 1985. Approximately half the papers presented at the conference were case studies, with most of the remainder being conceptual papers or reports on public participation. Also see McCallum (1986) and Munro, Bryant, and Matte-Baker (1986).

2. A systematic comparison of changes in the study's research design between the proposal and actual fieldwork can be found in the executive summary of our report to the National Science Foundation (Culhane, Friesema, and Beecher 1985).

3. These projects, deleting the two federal facilities (BARC and the Cummington radar dome) are the Newington Forest subdivision, Shepard Park residential development, Henrico County sewer system (whose controversial impact involved the development of rural eastern Henrico County), and the Yakima CBD project.

4. Our classification scheme is similar in several respects to that developed, completely independently, by the University of Aberdeen's Centre for Environmental Management. (See Bisset 1984: 470-471.)

5. Oral communication, faculty seminar, more years ago than we care to recall.

6
Project Impacts:
The Accuracy of Individual Forecasts

The purpose of this chapter is to describe representative cases audited in this study. Since the accuracy classification scheme described in Chapter 5 relies on evaluations of the match between an EIS's forecast and a project's impact, the best way to understand the scheme is to follow it through evaluations of a representative range of cases. Thus, this chapter allows the reader to examine the judgments represented in our audit classifications. We have also, of course, endeavored to select cases that are intrinsically interesting to readers familiar with federal resources projects and their environmental impacts.

This chapter contains two major sections. The first covers four EISs' projects, summarizing the forecasts, field data, and actual outcomes of the impacts audited during each project's fieldwork. These four projects' 38 cases include examples of 13 of the 17 primary accuracy classifications and all of the eight most important classifications. (See Chapter 5, page 142, for a definition of the audit's cases, forecast impacts.) The second section presents representative cases from the remaining 25 EISs, arranged from the most accurate auditing classifications to the least accurate.[1] Readers may examine either section, or both, depending on whether they prefer to evaluate accuracy of forecasts within the context of a specific project or in the more abstract order of our study's auditing system.

FOUR REPRESENTATIVE PROJECTS' IMPACTS

Each EIS within our sample possesses distinctive traits and, except for a few run-of-the-mill highway EISs, most present interesting examples of the quirks of environmental assessment. Four projects were selected to reflect the range of data available in the field sample. The first two, Paint Creek dam and Sequoyah hexaflouride plant, yielded some of the best impact data in the sample. The third, the Grand Teton National Park master plan, represents the middle range of the sample, in that it yielded some good and interesting cases, but also left some important unanswered questions. Finally, the Cleveland harbor diked disposal area project represents the bottom of the barrel--those situations in which no one possessed much, if any, valid information on a project's impacts. Collectively, these cases indicate the range of data available to audit EIS forecasts' accuracy--the good, the mediocre, and the ugly.

Paint Creek Dam

The Paint Creek dam was completed in June 1974. The EIS on this project was written on a project started before the passage of NEPA; in fact, the Paint Creek FEIS was filed in July 1974, after the dam was completed. The dam is located west of Chillicothe in southcentral Ohio.

The fieldworkers encountered good luck in auditing this particular EIS, obtaining satisfactory information on 11 of 13 candidate forecasts. As shown in Table 6.1, the field crew obtained enough data and information to evaluate 14 impacts--the largest number of cases per project in the sample. The Paint Creek data include both unusually good physiographic data and excellent time-series data, plus four impacts not on the candidate list that interviewees iden-tified as significant.

Flood Control. Flood control is the principal objective of Paint Creek dam. The EIS forecasts flood-control benefits in three ways: (a) the general statement that the dam "will alleviate the recurring, destructive floods downstream . . . along Paint Creek and Scioto River and will contribute to the reduction of flooding along the Ohio River," (b) a table of ex post facto estimates about reductions in flooding if the Paint Creek dam had been in place during six past major floods, and (c) a cost-benefit estimate of $1.83 million of annual flood-control benefits. Figure 6.1 graphs the monthly maximum daily streamflows, in cubic feet per second (cfs), from the Bourneville gauging station several miles downstream from the dam.[2]

148

Table 6.1
Summary: Paint Creek dam forecast impacts in field sample. Table includes both cases with adequate data and cases deleted because of inadequate data.

Category	I/M/O[a]	Impact Type	Primary Match[b]	Primary Data Type	Data Aptness
Phys.	Obj.	Flood reduction	Range–VF	TS	Adeq.
	Obj.	Water quality, downstream	Range–VF	Numeric	Low
	Impact	Instream min. flow	Close	TS	Exact
	Impact	Limnology	Close	TS	Exact
	Impact	Streambank erosion	Underant–	Verbal	Adeq.
	Impact	Sediment rate, reservoir	Underant–	Verbal	Adeq.
Biol.	Mit.	Fishery, monitoring	Close	Nominal	Adeq.
Econ.	Impact	Property tax base	Spurious	TS	Low
	Impact	Agricultural production	Less	TS	Exact
	Impact	Agr. prod., downstream	Underant+	Verbal	Adeq.
Social	Obj.	M&I water supply	Incons.	Verbal	Adeq.
	Obj.	Recreation visitation	Close	TS	Exact
	Impact	Police demand	Range–VF	Semi–TS	Adeq.
	Mit.	Archeology survey	Close	Nominal	Low

Subtotal = 14 impacts classified

Econ.	Impact	Flood damage reduction	(no avail. data)
	Impact	Employment, recr. business	(no avail. data)

Total = 16 forecast impacts sought during fieldwork

[a] Impact, mitigation, or objective.
[b] See Chapter 6 endnote 1, Table 7.1, or the text discussion of the particular impact for explanations of accuracy "match" codes.

The difficulty in auditing this forecast is to translate the relevant EIS passages into an operational forecast to be compared with postproject streamflow data. The EIS passage is not expressed as a forecast of future streamflow, even though Corps hydrologists undoubtedly calculated such maximum streamflows. Instead, the EIS presented estimates of hypothetical effects on past floods, which would make an understandable impression on readers. The section on pre-

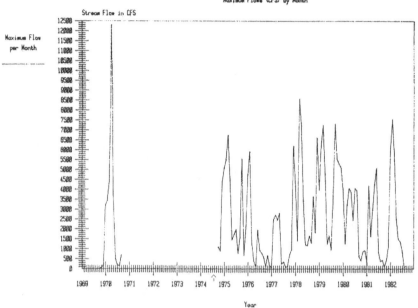

Paint Creek Dam and Lake: Stream Flow

Maximum Flows (CFS) By Month

Figure 6.1
Flood control: Peak monthly stream flows, Bourneville
gauge, Paint Creek, downstream of Paint Creek Dam, central
Ohio.   Forecast: Peak flow "regulation will alleviate the
recurring, destructive floods downstream from the dam along
Paint Creek and Scioto River, and will contribute to the
reduction of flooding along the Ohio River."   Dam
operational June 1974 (indicated by arrow).   Data for 1971-
1973 are missing.   Impact classified as "within the range of
vague forecast."

project hydrological conditions, however, contains various figures on the monthly average flows and flood stage frequencies during 41 preproject years. Flood levels range from a maximum recorded flood of 56,000 cfs (in 1964) down to the one-year flood stage of 9,000 cfs; the preproject monthly average maximum flow was 5,857 cfs (for Januaries).

If we translate the EIS's language into an operational forecast of a maximum postproject streamflow less than the nominal one-year flood of 9,000 cfs, the Paint Creek dam seems to have alleviated flood conditions. The annual maximum flow during the eight years after the dam's 1974 completion (marked by a small arrow in Figure 6.1) averaged 6,436 cfs, 28% less than the nominal one-year flood of 9,000 cfs. The highest maximum flow during the eight years after completion of the dam was 8,570 cfs, well below the preproject one-year, five-year (22,500 cfs), and ten-year (30,500 cfs) floods. The highest postproject monthly average maximum flow was 4,661 cfs (for Marchs) and the maximum flows for the peak-flow period of January through April averaged 4,063 cfs, at least a 20% reduction from the preproject maximum monthly average of 5,857 cfs.

So the Paint Creek flood-control forecast was classified as in the correct direction. However, given the imprecision of the forecast—that is, the vague "alleviate" phrasing and a hypothetical estimate about controlling past floods—the impact's primary classification was "within the range of a vague forecast." The actual outcome was also classified "as beneficial as forecast," in part because the actual streamflow maxima seemed consistent with the vague objective to "alleviate" floods. Moreover, field interviews indicated that Corps managers and other local informants believed the dam was accomplishing its flood-control objectives. According to Corps interviewees, operation of the Paint Creek dam was complicated by the operation of the nearby Rocky Flats dam.[3] The Ohio Department of Natural Resources (DNR) operates this dam to protect homesites on the Rocky Flats reservoir; thus, the state releases water into Paint Creek at precisely the most difficult period, without coordinating its releases with the Corps. Nonetheless, the Corps' dam operators appear to be alleviating maximum floods adequately.

Downstream Water Quality. A second Paint Creek objective involved water-quality improvements through "base flow stabilization." During the 1960s, congressional public works supporters argued that dams improved ambient water quality by providing a stable level of dilution for the pollutants in a watershed. Environmentalists do not subscribe to this notion of pollution control, but it nonetheless proved useful as another benefit that could be added to a dam's cost-benefit calculation. In the Paint Creek EIS, water-quality control is listed among the project's objectives (see quote below, in municipal water supply subsection), with $239,900 of annual benefits included in the cost-benefit computation.

Some limited water-quality data are available for years before and after the constuction of the Paint Creek dam. Since most of the watershed is farmland, nutrients pose the major pollution problem in Paint Creek. Phosphorus and nitrates were monitored occasionally at the Bourneville station from 1969 to 1976. The monitoring shows a decline in nitrates from a 1970-1972 average of 8.4 mg/l to a 1973-1976 average of 1.7 mg/l and a decline in phosphates from a 1969-1971 average of 0.44 mg/l to a 1973-1976 average of 0.12 mg/l.[4] These nutrient declines began in 1973, as the reservoir was being filled.

The Paint Creek water-quality impact was also classified as "within the range of a vague forecast." In one respect this forecast appears quite precise--$239,900 worth of annual benefits. However, the EIS reader has no idea about the relationship between the asserted dollar benefits and any measurable changes in physical water quality. In addition, the authors regard these alleged benefits as suspect, as undoubtedly many readers will, because the dam does not reduce total pollution but only redistributes the pollutants within the watershed.

In-stream Flows. As an operational constraint, a flood-control dam should maintain minimum downstream flows to provide adequate fish habitat. In addition, "low flow augmentation" is related to the water-quality benefits noted above, since the stable waterflow dilutes pollutants. Thus, the Paint Creek EIS forecast that the dam should be able to meet the schedule of monthly minimum flows shown in the first column of Table 6.2. Note that this forecast is the only one in the field sample that contains an explicit statement of the probability that the forecast outcome will occur: "these flows can be maintained nine years out of ten" (i.e., a .9 probability).

Minimum streamflow data from the Bourneville gauging station are shown in Table 6.2. Minimum flows fall below the EIS's schedule during only eight of the 96 months in the eight water years following completion of the Paint Creek dam. (A "water year" is calculated from October through

Table 6.2

Minimum in-stream flow, Paint Creek dam. Forecast: "Outflows will be provided as required to augment flows on Paint Creek in order to maintain the following minimum flows . . . : [schedule of minima below]. Based upon available records, these flows can be maintained nine years out of ten." Table shows monthly minimum flows, Bourneville gauge, Paint Creek, central Ohio. Non-measurement-error deviations below schedule minima printed in boldface. Dam operational June 1974. Forecast classified "Close."

Month	Schedule Minima	Minimum Flow in Water Year								
		69–70	74–75	75–76	76–77	77–78	78–79	79–80	80–81	81–82
October	41 cfs	24	78	119	71	134	64	368	89	**28**
November	35 cfs	39	176	142	116	147	327	407	251	104
December	33 cfs	95	520	342	65	386	457	510	281	124
January	33 cfs	180	1040	520	31	355	400	562	110	140
February	33 cfs	406	779	560	30	230	180	212	304	437
March	35 cfs	346	1130	341	380	230	288	416	201	489
April	37 cfs	582	266	109	239	436	275	302	158	292
May	43 cfs	380	205	73	135	500	191	256	697	144
June	49 cfs	167	222	79	71	229	191	222	120	145
July	50 cfs	57	78	128	50	60	70	239	147	**44**
August	49 cfs	40	62	76	65	47	160	307	64	**39**
September	46 cfs	29	61	60	56	56	145	97	43	**38**

September of the following year.) In four of these cases, the difference between the monthly minimum values and the schedule values is less than 4 cfs; these differences should not count as violations of the schedule since they are less than the error variance of the streamflow gauge, according to U.S. Geological Survey interviewees. The remaining four months' low flows average only a 9 cfs shortfall and all fall within the same 1981-1982 water year. In other words, the Corps met its in-stream flow schedule in seven of the eight water years of operation before our fieldwork and violated the schedule in only 4.2% of the months during that same period. Since ten years had not yet elapsed, we gave the Corps the benefit of the doubt and rated the forecast's "nine out of ten years" as close to the actual impact.

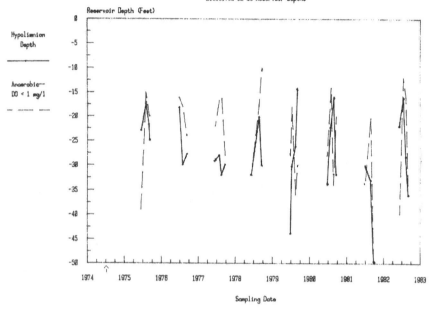

Paint Creek--Limnology
Dissolved O2 at Reservoir Depths

Figure 6.2
Limnological conditions, Paint Creek Reservoir, central
Ohio. Forecast: "The lake is expected to experience
thermal stratification during the warmer months, with
epilimnion ranging between 10 and 20 feet. . . . The high
organic load . . . is expected to produce anaerobic
conditions in the hypolimnion." Figure plots summer (June-
September) depths of the hypolimnion and depths at which
dissolved oxygen drops below 1.0 mg/l. Data derived from
monthly samples of temperature and DO at reservoir depth
increments taken by Corps personnel at the main sampling
location 400 feet upstream from dam. Full reservoir pool
achieved June 1974 (arrow). Impact classified as "close."

However, Geological Survey interviewees pointed out that the 1974-1982 period had recorded fairly wet years, so Paint Creek's record portended greater difficulty meeting its schedule during future droughts.

Reservoir Limnology. Warm-water reservoirs such as Paint Creek are commonly affected by eutrophication. The EIS passage on the issue thus predicted thermal stratification in the reservoir during warm months, a 10- to 20-foot-deep epilimnion (the upper stratum in the reservoir), and anaerobic conditions in the hypolimnion (the lowest stratum). The Corps kept detailed monitoring data on temperatures and disolved oxygen at 2- to 5-foot-depth increments at various sampling locations in the reservoir. This monitoring had been performed about monthly from 1975 to the present. In what was probably the most complicated data-manipulation task of the study (since the authors are political scientists, not limnologists) a BASIC program was written to estimate epilimnion and hypolimnion depths from raw temperature-depth readings.[5]

The relationship of hypolimnion depth (solid line) to the depth at which DO concentrations fall below 1.0 mg/l (dashed line) is shown in Figure 6.2. Anaerobic conditions exist in Paint Creek's summer hypolimnion strata, and often extend ten or more feet up into the thermocline (which was fairly indistinct on some monitoring days). Epilimnion depth (not shown in Figure 6.2) averaged 13.7 feet during 45 1974-1982 monitoring days and 12.7 feet during 32 summer monitoring days. Usually (21 of 32 monitorings) the epilimnion formed within the forecast summer range of 10-20 feet, though in eight monitorings on early summer days (early June into August) of 1979 and 1980, the Paint Creek epilimnion formed as shallow as four to eight feet.

In other words, as implied by the forecast, Paint Creek reservoir is indeed eutrophic. The basic tendency of warm-water reservoirs to become eutrophic was exacerbated in this case by the presence of the Greenfield municipal sewage treatment plant on Paint Creek north of the reservoir. The anoxic conditions in the Paint Creek thermocline confirm this, since they indicate a heavy biological oxygen demand. Thus, the Paint Creek limnology impact provides an example of a good, accurate forecast that received a "close" rating and an impact rated "as adverse as forecast."

Streambank Erosion. As noted in the methodology appendix, we interviewed local informants to identify any significant impacts that were well known to people familiar with the project. Local Corps officers and a downstream farmer stated that the major adverse impact of the dam had been an increase in downstream riverbank erosion. According to interviewees, the dam causes a constant flow in the river channel at a rate higher than the preproject average low flow. This significantly increases the "cutting" of banks by the channel current. The Corps' local dam managers claimed that the impact was unanticipated and the Corps did not know how to measure the phenomenon, so no data existed on the problem. Nonetheless, a consensus existed that the problem was extensive and serious.

The Paint Creek EIS did not contain a coherent prediction about this impact, only very general prose, such as,

> In addition to the nutrient loss, the water, without its sediment load, will have some slight degree of greater erosive power. Theoretically this increased erosional ability would increase the rate of meandering undercutting, and other changes due to stream motion. . . . . The pattern and mode of water retention and release over the years will be reflected in both physical and biological changes downstream . . . (Corps of Engineers 1974b: 74).

The coders did not code this vague passage as a forecast. Thus, the actual impact was classified as "underanticipated" and "more adverse" than implied by the EIS's non-forecast.

Reservoir Sedimentation. This case is very similar to the preceding one. Several interviewees, including the local Corps dam managers, indicated that sediment was filling Paint Creek reservoir at a significantly faster rate than expected. Higher than normal sedimentation could be expected at the Paint Creek dam because the watershed drained a 573-square-mile area of high runoff. The Corps had reportedly prepared a study of this problem at Paint Creek, but we were unable to obtain a copy of this document from the Corps during the fieldwork follow-up period.

The EIS's allusion to reservoir sedimentation is even more abstract than its passage about streambank erosion:

> The life expectancy of an impoundment depends upon several factors. If the barrier to flow, the dam, is constructed to achieve a long life, the limiting factor may well be that of sediment deposition. Very fine silts and clay particles may pass through the impoundment but most coarser sediments will be precipitated.

Unless some measures are provided for the removal of the accumulated sediments, the impoundment will eventually become filled with the stream coursing its way across the surface of the dam. The aquatic forms of life will change as the habitat changes and largely as a result of this change (Corps of Engineers 1974b: 74).

As with the previous case, our coders properly did not treat this (especially the second sentence) as a forecast. Thus, when interviewees identified the high sedimentation rate as significant, the impact was classified as "underanticipated" and "more adverse" than implied by the EIS's nonforecast.

Fishery Monitoring. The only satisfactory candidate biological forecast in the EIS involved monitoring (which is listed with mitigation measures under our procedures): "The new lake will be monitored through 1978 by Federal and State fisheries personnel and if supplemental stocking is needed, the stocking can be programmed at that time. If the program is successful, there should be a good fishery established within two or three years" (Corps 1974b: 65). An Ohio State University graduate student conducted this monitoring for his M.A. thesis (Claggett 1977). This case thus provides a good example of the straightforward nature of mitigation data: two interviewees, backed by the evidence of the thesis itself, confirmed that the monitoring had been done.
　　Claggett found that a fishery had been established naturally. Thus, even though stocking is normally required in new reservoirs, the Ohio DNR did not need to stock Paint Creek. The Ohio DNR stocked Paint Creek reservoir in 1982 with saugeye (a crossbred species), but strictly to enhance the reservoir's fishery for a fishing tournament. The fact that the monitoring was conducted by a graduate student, rather than by the federal and state personnel specified by the EIS, does not affect this case's classification because the local managers at the dam cooperated with the study. Since the monitoring was done, this mitigation forecast was classified as "close," and since the fishery was established without stocking, the ultimate outcome was rated "more beneficial" than implied by the forecast.

Property Tax Base. When a federal agency acquires land for a project, a tax-base loss is one of the most predictable short-term impacts of the proposal. EIS writers routinely balance their observations about such short-term tax losses with forecasts that economic development will increase long-term property tax receipts (via increased property values) and sales tax receipts. The Paint Creek EIS, for example, states: "Tax revenues formerly derived from project lands . . . will be lost and may cause an inconvenience to local governments until revenues begin to increase in response to economic activity induced by the project" (Corps 1974b: 75).

Property tax figures are normally the most reliable time-series data on project impacts, but the Paint Creek case is an exception. Paint Creek forms the boundary between Highland and Ross counties. Ross County tax records were kept in excellent condition, but reflected a systematic reappraisal for tax year 1977. (See Figure 6.3.) In 1974 Ohio enacted a constitutional amendment authorizing the appraisal of farmland at an agricultural-use value, rather than a market value for residential-commercial development. The Highland County property tax graph reflects the same general reappraisal, only for tax year 1976. Moreover, Highland County property taxes were terribly recorded, with pre-1974 aggregate tax receipts available only by collating individual receipts stored in shoeboxes.

Ohio DNR interviewees said that local residents often complained that the Corps' land acquisition had harmed the tax base of Greenfield School District in particular, but that insufficient economic development had occurred to offset this loss. However, the counties' tax data were too contaminated by the systematic reappraisal and Highland County's poor recordkeeping to verify this complaint. In Ross County, Paint Creek reservoir is located in western Paint Township (short-dash line in Figure 6.3) and the Paint Township valuations show a small drop in 1970, that is, during the reservoir's land acquisition phase. Downstream from the dam, Paint Creek flows along the southern boundary of Paint Township, through the northern half of Paxton Township, and then through Twin Township. Property valuations increase most in Paint and especially Paxton townships. These increases could reflect the value of flood protection downstream from the dam, with the lower increase in Twin Township attributable to the agricultural-use reappraisal (since Twin Township is just west of Chillicothe). However, we cannot know enough about the reappraisal's effects to reach a confident conclusion. In any case, the primary effect visible in both Highland and Ross counties' tax data is the 1976-1977 reappraisal. Thus, we classified this effect as "spurious"; that is, the change in appraisals soon after completion of the dam, rather than the dam's flood protection, caused the most clearcut changes in property valuation.

158

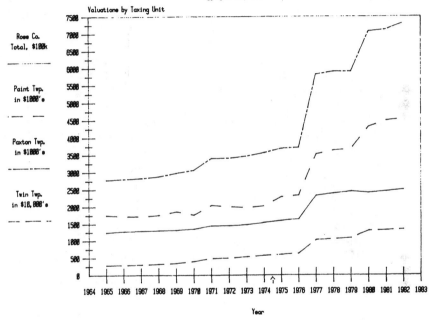

Paint Creek: Ross County Property Values
Aggregate & by District Valuations

Figure 6.3

Tax base: Property tax valuations, Ross County, Ohio, and townships adjacent to Paint Creek Reservoir. Forecast: "This loss of revenue is generally offset by increases in the tax base generated by project-induced activity." Land acquisition began January 1966; reservoir operational June 1974 (arrow). Impact classified as "spurious."

159

Agricultural Productivity. Since reservoir land acquisition principally involved farmland, the EIS also forecast a decline in agricultural productivity: "The Soil Conservation Service estimates an annual loss of agricultural income due to the project at $825,000, over 3.4 percent of Highland County's agricultural income" (Corps 1974b: 51). The Department of Agriculture collects detailed farm production data on both acreages planted and cash sales. According to both local USDA agents and farmers, the primary crops in Paint Creek bottom lands are corn and soybeans.

As shown in Figure 6.4, Highland County corn and soybean receipts fell by $1.8 million between 1974 and 1975, with Highland County falling behind Ross County's production of the same crops. However, it is unclear that this effect can be attributed to the Paint Creek dam, since the reservoir reached full pool in June 1974, well before the 1974 harvest. The decline between 1973, the last full year before completion of the dam, and 1975 was only $217,000, and Highland County corn and soybean production actually increased by $1.6 million from 1973 to 1974. Moreover, by 1978 Highland County had regained its lead over Ross County. Farm commodities are subject to volatility in prices and substitution among commodities, of course, and the dominant pattern in Figure 6.5 is a strong secular trend. Since the impact is, at worst, small and short-term, it was classified "less than forecast."

Downstream Agricultural Production. Three interviewees—a local Corps manager, a farmer, and the leading farm informant in the Ross County business community—argued that the Paint Creek dam significantly decreased the risk of flooding within the floodplain downstream in Ross County. This decreased risk allowed regular planting and harvesting on these lands. The Corps interviewee argued that the market value of these protected downstream acres increased from $50/acre to $1000/acre. However, our township- and county-level data do not conclusively support this. As noted above, land values in downstream Paxton Township increased more than in Ross County as a whole. On the other hand, the numbers of acres harvested in downstream Ross and upstream Highland counties both increased sharply and at the same rate in 1977, but the Ross County harvest acreage graph flattened out during 1978-1980, while the Highland County acreage kept increasing. In other words, insofar as this asserted flood control benefit occurred, it seems to have been fairly localized, not county-wide.

160

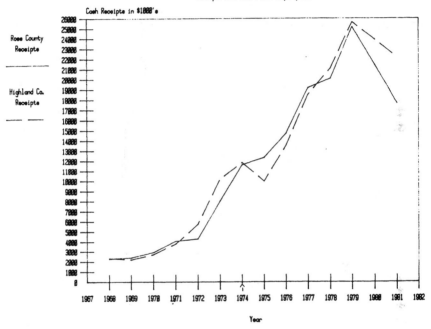

Paint Creek Area: Agricultural Effects
Receipts from Corn & Bean Crops by Co.

Cash Receipts in $1000's

Ross County Receipts

Highland Co. Receipts

1967 1968 1969 1970 1971 1972 1973 1974 1975 1976 1977 1978 1979 1980 1981 1982

Year

Figure 6.4
Agricultural productivity: Cash receipts from corn and
soybean crops, Ross and Highland counties, central Ohio.
Forecast: "[E]stimate an annual loss of agricultural income
due to the project at $825,000, over 3.4 percent of Highland
County's agricultural income." Paint Creek reservoir
filled, precluding farm uses, in 1974 (arrow). Impact
classified as "less than forecast."

Corps interviewees claimed that these flood control benefits were unanticipated. Frankly, we regard such benefits to be standard for such a project, just as taking upstream farmland out of production is a standard cost of a reservoir. Nonetheless, in describing flood control benefits of $1.83 million, the EIS mentions only the following types of flood damage: "death, injury and illness," "social and economic insecurity," "the necessity to commit labor and material to emergency and relief operations," and "aesthetic degradation" (Corps 1974b: 50). Thus, this alleged benefit was classified as "underantici-pated-beneficial."

Municipal Water Supply. The Paint Creek EIS lists municipal water supply as an objective of the project, but qualifies this listing:

> The project will serve the purposes of flood control, general recreation, fish and wildlife recreation, water supply, and water quality control. . . . [S]tudies . . . as published in House Document No. 587, 87th Congress, 2nd Session, modified the scope of the project as authorized in the 1938 legislation to include recreation. Approval to include potential water supply and water quality control was granted by the Chairman of the Subcommittee on Public Works of the Senate Appropriations Committee in March 1970. The project has a benefit-to-cost ratio of 2.0 to 1. . . . Water supply is available for future needs. Although the benefits for water supply were computed to be $9,700 annually (1973 prices), they were not included in the benefit-cost ratio (Corps 1974b: 1,50).

This passage, especially the third sentence, portrays water supply as a bona fide project purpose. However, the EIS qualifies this assertion by referring to "potential water supply" that is only "available for future needs," and by not including these benefits in the project's B/C ratio.

Local interviewees clarified this ambiguity by explaining that Paint Creek reservoir was never really intended to provide a municipal water supply to Greenfield. The original reservoir was to have had a design elevation at 785 feet above sea level. The elevation was raised to 787.5 feet, and the increase was justified as providing potential water supply for Greenfield. The increased elevation actually maintains year-round access to the reservoir at the Greenfield marina, thus benefiting tourism in Greenfield.

Moreover, the higher pool elevation could have caused flooding at the Greenfield sewage treatment plant. To prevent such flooding, with its attendant pollution, the Corps installed a flood protection dike and two flood pumps at the sewage plant. Greenfield has never used the reservoir water supply potential, relying instead on its perfectly adequate wells. We classified this outcome "inconsistent" with and "less beneficial" than implied by the EIS's language.

Recreation Visitation. The Paint Creek EIS's principal recreation forecast, with two postproject numbers, qualifies as one of the few time-series forecasts in the field sample. (See Table 5.5.) The Corps (1974b: 50) estimated "an initial visitation of 250,000 and a probable ultimate visitation of 575,000." Recreational use is an authorized objective of the project, contributing $323,700 of annual benefits to the dam's B/C computation.

Figure 6.5 graphs the official recreation use of the Paint Creek facility in visitor-days. Recreation data must be read with a critical eye. The Corps and the Ohio DNR maintain traffic counters at all entrance roads to the reservoir. The Corps calculates recreation visitor-days from counter readings using a simple formula: the number of counter increments, divided by two (an "ingress" and an "egress" per vehicle), multiplied by an assumed 3.5 visitors per vehicle equals the number of visitor-days. The state data in Figure 6.5 are based on exactly the same traffic counter readings as the Corps data, but the Ohio DNR changed its formula during the period (allegedly because visitation was critical in Ohio state parks' budgeting). In any case, the Corps' figures indicate that the EIS's recreation forecast was accurate. In 1975, the first full year of recreation after the reservoir was completed, annual visitation totaled 328,900 visitor-days. In 1979 visitation reached 541,100, and in 1982 hit 767,125. The 1982 record exceeded the EIS's long-run visitation forecast because of a fishing tournament held at Paint Creek reservoir. Nonetheless, the 1978-1982 five-year moving average of 550,900 was close to the EIS's target. Thus, this outcome was classified as "close" to the forecast objective.

Park Police Services. Among the secondary impacts of recreational use of Paint Creek reservoir, the Corps (1974b: 52) predicted, "Demand for government services, for example, fire and police protection, will probably increase. . . ." Later in the EIS (p. 72) the Corps referred to the demand as

163

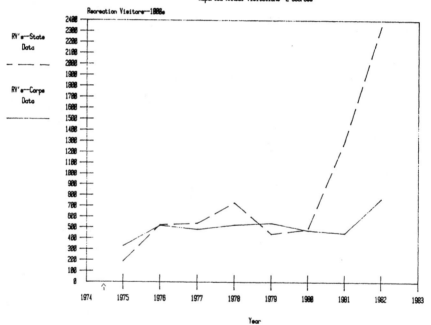

Paint Creek Dam: Recreation Visitation
Reported Annual Visitations—2 Sources

Figure 6.5
Recreation visitation, Paint Creek Reservoir and Paint Creek
State Park, central Ohio.   Forecast: "Estimates made in
1965 indicate an initial visitation of 250,000 and a
probable ultimate visitation of 575,000."   First recreation
facilities at reservoir opened in May 1974 (arrow).   Impact
classified as "close."

164

"a long-term and gradual need for expansion of fire and police protection." The forecast implies that the demand will affect local government, such as the county sheriff's office. Such references to increased public-service demands are a common social forecast in EISs.

The Ohio DNR increased the number of state park officers with police powers from none in 1974 to two in 1975-1976, three in 1977-1978, four in 1979, five in 1980-1981, and six in 1982. Ohio DNR interviewees indicated, however, that Paint Creek experienced only the small number of petty crimes that are normal in any state park. The field crews could not find any evidence of a significant park-related increase in regular police activity, such as increased sheriff's calls. The increase in park officers with police powers seems consistent with the Corps' "gradual need" forecast, but since the EIS is not explicit about what level of government will meet police demands, we classify this impact as "within the range of a vague forecast."

Archeological Survey. Paint Creek is located in an area containing many Native American burial mounds. One site, Plum Run Mound, is listed on the national register and lay within the Paint Creek floodstage pool. Thus, as a mitigation, the Corps (1974b: 55) noted, "Because prolonged inundation could be detrimental to the scientific value of the mound, the Ohio Historical Society, with funds made available by the National Park Service, is scheduled to undertake excavation of the mound after Fiscal Year 1974."

The Paint Creek State Park manager confirmed that the Plum Run Mound archeological survey had been completed and referred us to the survey's author, Professor Bobbi of Ohio State University. The staff of the Park Service's Mound City Group National Monument in Chillicothe could not provide a copy of the survey, but also referred us to Professor Bobbi. Unfortunately, Professor Bobbi died the week before our fieldwork, so we could not obtain a copy of the survey from him. (Our information on this mitigation is thus rated only "low adequate.") Nonetheless, we have sufficient confidence that the survey was actually completed to classify this mitigation forecast as correct.

Sequoyah Uranium Hexafluoride Plant

The Sequoyah plant provides a second set of cases for which very good monitoring data were available. Uranium hexafluoride is produced as an intermediate step in the

Table 6.3
Summary: Sequoyah uranium hexafluoride plant forecast impacts in the field sample. Table includes both cases with adequate data and cases deleted because of inadequate data.

Category	I/M/O[a]	Impact Type		Primary Match[b]	Primary Data Type	Data Aptness
Phys.	Impact	Radiation, ambient, air		Exceeds	TS	Exact
	Impact[c]	Air quality, fluorides		Incons.	TS	Exact
	Impact[c]	Groundwater, raffinate		Range-VF	TS	Exact
	Mit.	Raffinate waste management		Complex+	Nom.	Adeq.
	Mit.	Air quality, NO_x		Unant.+	Nom.	Adeq.
Biol.	Impact	Benthic effects[x]		NIm/NFor	Exper.	Exact
	Mit.	Benthic study		Close	Nom.	Exact
Econ.	Obj.	Mfg. process diversity		Range-VF	Verbal	Adeq.
	Impact	Employment, plant		Close	TS	Exact
	Impact	Property taxes		Close	TS	Exact
Social	Impact	Traffic volume		Range-VF	Numeric	Low
Subtotal = 11 impacts classified						
Phys.	Impact	Water quality, N and F	(no available adequate data)			
	Mit.	Air & water sampling, misc.	(interview info. inadequate)			
	Mit.	Fluorides control measures	(interview info. inadequate)			
Social	Impact	Recreation use, region	(data error-ridden)			
Total = 15 forecast impacts sought during fieldwork						

a Impact, mitigation, or objective.
b See Chapter 6 endnote 1, Table 7.1, or the text discussion of the particular impact for explanations of accuracy "match" codes.
c Controversial impacts.

manufacturing phase of the uranium fuel cycle. The Kerr-McGee Corporation's Sequoyah plant, located southeast of Muskogee in eastern Oklahoma, is one of two UF_6 plants in the U.S. It makes UF_6 from uranium "yellowcake," which it then ships to one of three gaseous diffusion plants (UF_6 is the gas diffused) that enrich the concentration of U_{235} for use in reactors and bombs. The Sequoyah plant is not the Kerr-McGee facility involved in the Karen Silkwood affair; Ms. Silkwood worked at Kerr-McGee's plant in Crescent, 50 miles north of Oklahoma City. The Sequoyah plant, however, was the site of a nationally publicized incident. In January 1986, a 14-ton UF_6 shipping cylinder ruptured after it had been mishandled, killing one worker and injuring 36. UF_6 reacts violently with moisture in the atmosphere, forming hydrofluoric acid. Hydrofluoric acid trauma, not radiation exposure, caused the death and most injuries in the incident.[6]

Notwithstanding this incident, the Sequoyah plant has been closely regulated by the AEC and its successor, the NRC. The plant received its original AEC license and went into operation in February 1970. The 1975 FEIS in the sample covered the relicensing of the plant, which was a procedural device to justify a "grandfather" EIS; the license was renewed in 1977. AEC/NRC regulations require licensees to conduct extensive monitoring of plant conditions and impacts. Thus, as shown in Table 6.3, the field crews were able to obtain excellent data on 11 candidate forecasts, including three physiographic impacts with up to four fine monthly time-series per impact, plus a good scientific-design benthic study.

Ambient Radiation. Radioactive emissions are naturally the primary concern of the NRC's EIS writers. Based on a series of assumptions and a dispersion model, the EIS included a table of maximum groundlevel uranium concentrations at various points within a four-mile radius of the plant. The maximum off-site forecast concentrations, at State Highway 10 just east of the plant's fenceline, were shown as 1.56×10^{-15} microcuries per milliliter (uCi/ml) of soluble uranium and 1.87×10^{-14} uCi/ml of insoluble uranium. The EIS noted that these amounts "correspond to 0.05% and less than 1.0% of the allowable limits . . . as specified in . . . 10 CFR 20" (NRC 1975: V-17)—that is, the limits specified by applicable NRC regulations. The NRC required Kerr-McGee to monitor ambient radiation emissions. Kerr-McGee recorded its monitoring data as a "proportion MCP" (i.e., as a percentage of the 10 CFR 20 limit of 5×10^{-12} uCi/ml).

The Sequoyah plant's actual monthly average ambient uranium concentrations exceeded the forecast maximum of 0.4% MCP by an order of magnitude. (That is, 1.87×10^{-14} plus 1.56×10^{-15} equals 0.02×10^{-12}, which equals 0.4% of the 10 CFR 20 limit of 5×10^{-12} uCi/ml total uranium.) From 1975 through 1982, the average monthly readings across all four plant fencelines fluctuated between 10% and 50% MCP, as shown in Figure 6.6. The monthly readings at the east

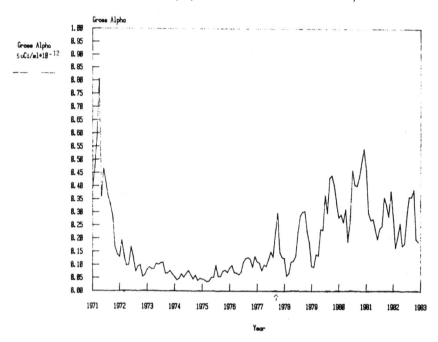

Sequoyah UF6 Plant: Air Quality

Figure 6.6
Radiation: Ambient airborne uranium concentrations,
Sequoyah UF$_6$ Plant, Oklahoma. Forecast: "The maximum off-
site ground level uranium concentration . . . calculated
values amount to about one percent of the 10 CFR 20
allowable limit for insoluble airborne uranium and about
0.05 percent of the permissible limit for soluble uranium"
(i.e., a total maximum concentration of 2.0 x 10^{-14} uCi/ml.
Figure plots monthly average plant fenceline concentrations,
expressed as a proportion of the "10 CFR 20 limit" of
5 x 10^{-12} uCi/ml.   The plant began operation in February
1970; after the 1975 FEIS, it was relicensed in October 1977
(arrow).   Accuracy classification: "impact exceeds
forecast."

fenceline, next to Highway 10, normally fluctuated between 3% and 48% MCP. However, all monthly averages fell well below the 10 CFR 20 limit, with a maximum monthly average across all four fencelines of 54% MCP and a maximum east-fenceline monthly average of 86% MCP.

In general, impacts that were significantly greater or less than forecast but still within some bounds noted in the forecast did not receive the most inaccurate classification. The Sequoyah radiation impact was in the "correct" direction because the actual impact fell below the forecast's "10 CFR 20" limit. Radiation emissions, however, were classified as "impact exceeds forecast."

Fluoride Emissions. In another ambient air quality forecast, the NRC staff predicted that average off-site ambient fluoride concentrations would not exceed 0.070 ug/m^3, with a maximum annual average at the plant's fence-line of 0.337 ug/m^3. The health effects of environmental fluorine have, of course, been debated off and on for decades. At the time of the EIS there were no federal fluoride regulations, so the NRC compared these concentrations to the only existing standard, the State of Washington's 0.5 ug/m^3 (a conservative standard probably designed to protect forest growth).

Figure 6.7 shows the ambient monthly mean fluorine concentrations averaged across all four fencelines, which Kerr-McGee monitored in ug/l (1 ug/l = 1000 ug/m^3). Fluorine concentrations were one to two orders of magnitude larger than the forecast value of 0.07 ug/m^3. Moreover, Sequoyah's fluorine concentrations also exceeded the EIS's reference "Washington State" standard in 71 of the 96 months from the date of the FEIS through 1982, with an October 1979 monthly high of 10.7 ug/m^3, or 21 times that standard. The impact is therefore classified as "inconsistent" with the forecast, in the "wrong" direction (i.e., above the 0.5 ug/m^3 standard), and "more adverse" than forecast.

The reason for this wrong forecast, which is the largest among the "inconsistent" impacts in the sample, is that national standards were promulgated after the FEIS. The U.S. Environmental Protection Agency issued state implementation plan guidelines in 1977-1978 for the aluminum and fertilizer industries, the only industries subject to fluoride regulation under the Clean Air Act. Those guidelines suggested performance standards that would result in offsite concentrations of about 8.0 ug/m^3. Note that Kerr-McGee is not actually governed by this standard: a UF_6 plant is not in one of the two industries subject to the

Sequoyah UF6 Plant: Air Quality

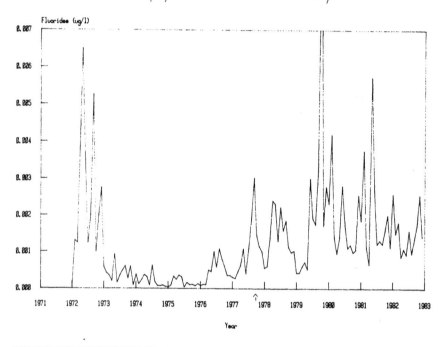

Figure 6.7
Air quality: Ambient fluoride concentrations, Sequoyah UF$_6$
Plant, Oklahoma.   Forecast: "Annual average fluoride
concentration is not expected to exceed 0.070 ug/m^3 . . .
maximum annual average . . . may reach 0.337 ug/m^3, which
is less than the Washington state standard of 0.5 ug/m^3."
Figure plots monthly monitored fluoride concentrations,
averaged across the four plant fencelines, in ug/1 (1 ug/1 =
.001 ug/1).   The Sequoyah plant began operations in
February 1970; it was relicensed in October 1977 (arrow).
Impact classified as "inconsistent."

guidelines, Oklahoma has not promulgated fluoride regulations, and the standard was a secondary, not a primary or
health-related, standard. Though not bound by the new
guidelines, Kerr-McGee kept its fluoride emissions below the
guideline level, since all readings except for the deviant
month of October 1979 fall below 8.0 ug/m^3. Nonetheless,
Figure 6.8 shows Kerr-McGee allowed its fluoride emissions
to increase up to the new standard. We assume the company
allowed this increase to avoid the pollution control costs
of complying with an old, superseded standard.

Groundwater Seepage, Raffinate Pond. The Sequoyah
plant's main operating problem during the early 1970s
involved leaks into the groundwater from the plant's
settling ponds for process liquids or "raffinate." Kerr-
McGee operates a set of monitoring wells to detect such
seepage. Despite the EIS's extensive, highly technical
discussion of raffinate monitoring and mitigation measures,
the key forecast was quite vague: "On the basis of these
findings, . . . the staff has concluded that continued
temporary storage of raffinate in the existing ponds subject
to the conditions described [earlier in the FEIS] does not
represent an appreciable risk of contamination of the area
water resources" (NRC 1975: V-7).

On the basis of interviews with plant officials and a
tour of the plant, we confirmed that Kerr-McGee had replaced
the raffinate ponds that had caused groundwater problems in
the early 1970s. Essentially, the main raffinate pond had
been replaced with a four-pond configuration in which the
first pond was used for extracting radioactive solids and
the remaining three for progressively clarifying the raffinate liquids. These ponds had been lined with a double
layer of high-strength hard rubber.

Kerr-McGee monitoring wells provided monthly groundwater
data on uranium, nitrate, and fluoride concentrations, plus
gross alpha and beta radioactivity. Because of the vagueness of the forecast, our first task in interpreting these
data was to concoct an operational definition of the forecast. On the basis of a portion of the FEIS discussion that
implied that recent uranium readings of 0.25 mg/l (which
equals 250 ug/l) and nitrate concentrations of 600 mg/l were
acceptable, we used those values as estimates of "no appreciable risk" levels. For the other monitored pollutants, we
examined time-series charts looking for levels lower than
those of the problematic pre-FEIS levels.

Uranium concentrations in the raffinate monitoring well
are graphed in Figure 6.8. Except for one bad post-FEIS

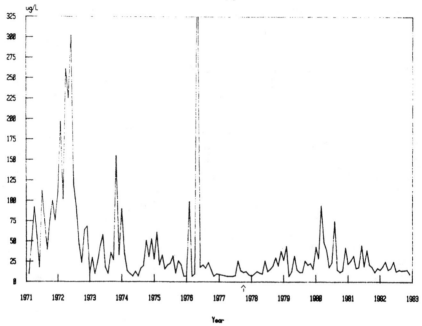

Sequoyah UF6 Plant: Groundwater Quality

Raffinate Pond Seepage, Uranium

Figure 6.8
Groundwater quality: Seepage of uranium (in ug/1),
raffinate pond monitoring well #2303, Sequoyah UF$_6$ Plant,
Oklahoma.   Forecast: "[C]ontinued temporary storage of
raffinate in the existing ponds subject to the conditions
[monitoring] . . . does not represent an appreciable risk of
contamination of the area water resources."   The Sequoyah
plant began operations in February 1970; it was relicensed
in October 1977.   Impact classified as "within the range of
vague forecast."

monthly reading, at 428 ug/l, all the readings fell under 100 ug/l, mostly under 50 ug/l. Nitrate seepage was recorded as less than 100 mg/l most months, except for a problem period from December 1975 through June 1976 when seepage levels reached 461 mg/l. Even that, however, still falls below our operational forecast of 600 mg/l. Especially after 1976, fluoride levels generally fluctuated around 0.5 mg/l, compared with problem episodes of 8.9 mg/l in 1971 and around 2 mg/l in 1972-1973. Gross alpha and beta readings were also stable, with almost all monthly readings less than 50 picocuries/liter, compared with the problem episodes during 1972 when levels fluctuated between 110 and 1158 pCi/l.

This case thus involves some of the most extensive information in the study—four fine monthly time-series data sets, supplemented by extensive interview information and a site tour. As a secondary classification, this case receives a "mitigation done" code because of the improved raffinate pond lining. Monitoring of raffinate seepage, indicating improvements following the FEIS, confirms that the impact is in the forecast's "correct" direction and suggests that the impact is "less adverse" than implied by the forecast. However, the impact's primary classification is "within the range of a vague forecast" because the forecast was imprecise.

Raffinate Waste Management. The FEIS's alternatives chapter reviewed several methods for disposing of the plant's raffinate liquids. The fourth alternative involved chemical precipitation of the radioactive elements in the raffinate. After recovery of the uranium, thorium, and radium, the residual ammonium nitrate could be used as a fertilizer. The NRC staff noted, however, "Considerable additional information must be developed before [this process] can be considered as an acceptable solution to the raffinate disposal problem." The fifth alternative consisted of two processes for chemically separating radioactive elements and either recycling nitric acid or emitting nitrogen as an air pollutant. The NRC concluded the alternatives chapter by stating, "it is the considered opinion of the Staff . . . that [Kerr-McGee] should be directed toward the development and implementation of one of the two proposed processes which include conversion of the radionuclides to solids as well as decomposition of nitrates," that is, alternative 5 (NRC 1975: IX-7 - IX-8). Raffinate waste disposal was the most controversial Sequoyah issue at the time of both the EIS and the fieldwork.

Kerr-McGee, in fact, pursued alternative 4, the system that produces ammonium nitrate as a byproduct. The company was also seeking a patent for its process to recycle raffinate solids as a uranium mill feedstock. The company developed these systems largely for business reasons, including its contractual obligation to account for uranium to its customers, the capital cost of alternative 5, and the potential for selling the ammonium nitrate.

Kerr-McGee knew it faced a marketing problem in commercializing its raffinate fertilizer process. It conducted a series of experiments on the safety of its recycled ammonium nitrate. Plant officials found that the test plots fertilized with recycled raffinate contained lower uranium concentrations than plots treated with commercial fertilizer. (Some fertilizer contains high concentrations of uranium. Indeed, fertilizer manufacturers supplied yellowcake feedstock to the Sequoyah plant.) Notwithstanding these results, local farmers understandibly resisted buying fertilizer recycled from a nuclear plant.

The classification of this impact is complicated. On the one hand, Kerr-McGee chose to implement alternative 4, instead of the alternative supported by NRC. On the other hand, both alternatives 4 and 5 proposed processes for recycling raffinate, and Kerr-McGee had developed a recycling system for both raffinate solids and the residual liquids. Company interviewees argued their system was economically superior to alternative 5, and we suspect the company's system was environmentally preferable to the one advocated by NRC. Thus, especially since the NRC phrased its forecast about this mitigation as only a suggestion, we classified this case as "complex but essentially accurate."

NO_x Mitigation. Kerr-McGee interviewees noticed on our list of candidate forecasts an NO_x entry (which we did not audit since we obtained good data on the physiographic cases discussed above) and volunteered information about a mitigation measure that had not been anticipated by NRC's EIS writers. According to the plant managers, the plant's NO_2 emissions were pale yellow and, since the plant processed uranium "yellowcake," these emissions caused a "public relations problem." Even though the plant's emissions were well within Oklahoma's air quality standards, the company decided to install a $300,000 NO_x control system. Since the EIS had characterized NO_x impacts as insignificant, and thus had not contemplated mitigation measures, this outcome was classified as "unanticipated but essentially beneficial."

174

Benthic Effects and Monitoring. In one of the few biological forecasts in the EIS, the NRC concluded that the Sequoyah plant's wastewater effluents would not affect the adjacent Illinois River: "It appears highly probable that the required dilution would be readily attained quite close to the plant outfall and that there would be little, if any, adverse effects of the liquid waste stream on either the microscopic or macroscopic river biota" (NRC 1975: V-8). Nonetheless, later in the EIS, the NRC staff concluded that the plant's environmental monitoring program should include a study to confirm that the plant was not adversely affecting benthic populations in the Illinois River.

Kerr-McGee hired an Oklahoma State University biologist to conduct a four-year study of benthic conditions in the Illinois River. The field team obtained a copy of that study, so the mitigation was classified as accurate.

The study found benthic conditions to be better at the plant's outfall than either upstream or downstream (Russell 1982). Shannon species diversity index readings at the plant's effluent outfall averaged 3.22, compared with 2.95 upstream and 2.40 downstream. Russell persuasively argued that the lower downstream benthic readings were the spurious result of intrusion of salty water into the Illinois River from the Robert Kerr Reservoir. Given the higher readings at the plant outfall (probably also an artifact of peculiar conditions), he concluded that the plant did not actually harm benthic conditions in the river. Since the observed differences in benthic conditions were not significant and probably spurious, this case was classified "no clear impact, no significant impact predicted." The key facet of this classification is that the impact, if any, is not clear because of the peculiar conditions in the Illinois River.

Objective: Manufacturing Process Diversity. The NRC (1975: IX-1) described the primary specific benefit of the Sequoyah plant as diversification of the uranium hexafluoride manufacturing process: "[An] added benefit of the Sequoyah facility is that it uses a process which is more readily capable of accommodating high sodium content uranium concentrate feed than the competing hydrofluor process." Later on the same page, the EIS writers added a sweeping conclusion to the benefits section: "It is fully anticipated that essentially all of these benefits will continue to accrue during the projected life of the plant at least at the current rate and at some accelerated rate if the production capacity is increased in the future." This sentence qualifies this description of objectives as a

175

forecast, albeit a very imprecise verbal one.

A small proportion of the Sequoyah plant's feedstock is indeed nonstandard liquid feedstock.[7] This feature of the plant provided little advantage to Kerr-McGee, according to plant managers, and it seems to represent a fairly subjective benefit to the Department of Energy and the nuclear industry generally. Nonetheless, the plant actually uses nonstandard feedstocks, so this very ordinary objective was classified as "within the range of a vague forecast."

Plant Employment. The same benefits analysis noted that the Sequoyah plant employed "more than one hundred employees" with an annual payroll of $1.25 million. As with the preceding case, this factual statement qualifies as a forecast because of the generic forecast, "all these benefits will continue to accrue . . . at least at the current rate," at the end of the EIS's benefits section.

Sequoyah's employment was never less than 100 from the plant's 1970 opening through 1982. Employment and the plant's payroll increased steadily from lows of 102 employees and $955,000 in 1972, to 111 employees and $1.4 million in 1976, to 162 employees and over $3 million by 1981, before decreased demand for uranium fuel led to layoffs of a dozen employees in 1982. (At the time of our site visit, a huge inventory of uranium yellowcake was stored on the plant grounds.) The Sequoyah plant accounted for 19.6% of its county's manufacturing employment during the 1970s. Since the forecast essentially established minimum values of 100 employees and $1.25 million payroll, this impact was thus classified as "close" and "as beneficial as forecast."

Property Taxes. A third economic benefit mentioned in the same section and forecast to "continue to accrue" is annual county tax payments of $100,000. Kerr-McGee's taxes on the Sequoyah plant were about $95,000 during 1971-1974, then increased from $106,605 in 1975 to $169,489 in 1982 (with an anomalous drop to $104,474 in 1979). The plant is a major tax source in Sequoyah County. Since all annual tax payments after the 1975 FEIS exceeded $100,000, this forecast was rated as "close" to the actual impact. This case is, of course, very similar to the preceding one. It is very easy to predict simple economic outcomes like taxes and employment for a completed plant with known assessments and operating experience, especially given the conservative "at least at the current rate" form of the forecast.

Traffic Volume. The Sequoyah EIS differs from the average EIS in the field sample in that it contains almost no coverage of social impacts. The EIS contained a vague, no-impact forecast about traffic: "The total increase in truck traffic of about 27 inbound shipments and about 8 outbound shipments per week is not expected to have any measurable effect on . . . the traffic pattern in the area," and "located in an area of high recreational activity, impact of the plant on movement of traffic is not discernible" (NRC 1975: III-2, X-11). The FEIS predicted no archeological or historic impacts because there were no national register sites nearby.[8] The only data bearing on a "no impact" recreation forecast proved to be invalid because of frequent changes in methods of counting recreation visits. Since our fieldwork protocol required us to audit a social forecast, we chose the traffic forecast by default. The available traffic data is also not very good—six average daily traffic counts between 1976 and 1983 on Highway 10 at the plant—but we rated it minimally adequate to reach a conclusion about the forecast.

The data show Highway 10 traffic has fluctuated between 325 and 450 vehicles during the post-FEIS period. This small state road opened at the same time as the Sequoyah plant, so there was no preproject traffic. With 111 to 162 employees commuting to work, plus some truck traffic, and no other businesses or residences located on this short road, the Sequoyah plant clearly accounts for most of the Highway 10 traffic. Just as clearly, Highway 10's ADT of about 400 approximates the random daily variance in the 6,000 ADT on Interstate 40, the major east-west highway located a mile south of the plant. The Sequoyah plant's traffic thus seems to be both "within the range" of the vague, "traffic pattern" forecast and quite trivial.

Grand Teton National Park Master Plan

The Grand Teton master plan illustrates some of the data-gathering difficulties the field crews experienced in auditing the average project in the sample. Grand Teton is a major park in western Wyoming, immediately south of Yellowstone. Various elements of the master plan were implemented from the March 1976 approval of the plan through the fieldwork. (The master plan also restated the conclusion of the 1974 Jackson airport FEIS, which is a separate EIS in the field sample.) The fieldworkers acquired some solid data and nonquantitative interview information. How-

177

Table 6.4
Summary: Grand Teton National Park master plan's forecast impacts in the field sample. Table includes both cases with adequate data and cases deleted because of inadequate data.

Category	I/M/O[a]	Impact Type	Primary Match[b]	Primary Data Type	Data Aptness
Social	Obj.	Wilderness designation	Complex+	Nom.	Exact
	Impact	Residential displacement	Less	Verbal	Adeq.
	Impact	Recreation visitation	Range—VF	TS	Exact
Phys.	Obj.	Sewage treatment	Close	Nom.	Adeq.
	Obj.	Fire management	Close	Nom.	Exact
	Impact	Forest fires	Not yet	TS	Exact
Econ.	Impact	Concessionaire business	Less	TS	Exact
	Impact	Property tax base	NIm/Some	Verbal	Adeq.
Subtotal = 8 impacts classified					
Phys.	Impact	Auto pollutant emissions	(no adequate data available)		
Econ.	Impact	Property values	(no adequate data available)		
Social	Impact	Traffic, outside N.P.	(no adequate data available)		
	Obj.	Preserve distinctive ecological conditions	(forecast unevaluatable)		
Total = 12 forecast impacts sought during fieldwork					

[a] Impact, mitigation, or objective.
[b] See Chapter 6 endnote 1, Table 7.1, or the text discussion of the particular impact for explanations of accuracy "match" codes.

ever, data on an equal number of important impacts were inadequate or disappointing. The crews had to accept interview information, rather than solid water-quality data, on sewage treatment problems within the park. The only data available on air pollution were for a monitoring station in an airshed 20 miles away from the park. (This situation was common for air-quality data.) The EIS did not contain any normal biological forecasts that could be audited, which we found surprising for an EIS produced by an agency traditionally managed by ranger-naturalists.

The 1975 Grand Teton master plan and FEIS, in line with National Park Service policy at the time, emphasized park preservation over recreational uses. The principal objective of the plan was phrased in abstract terms: "Elimination of intrusive cultural features [from within the park] will enhance the area's ecological and esthetic values through a gradual return to a vignette of primitive America" and "The principal 'product' of Grand Teton National Park is a dynamic equilibrium of the area's biotic and abiotic resources that are not modified by man and his works. The significant benefits from this type of management are esthetic and scientific, and thus are not readily quantifiable" (NPS 1975: 29, 38). These statements aptly characterize NPS policy; the "vignette of primitive America" phrase, for example, simply repeats the key theme of park preservation from an influential 1963 report on park policies by A. Starker Leopold's advisory committee to Secretary of the Interior Udall (Chase 1986: 34). That policy is controversal among both developers and some environmentalists (e.g., Chase 1986). We believe the Park Service implemented that policy in Grand Teton, but we quickly concluded that an objective couched as "a dynamic equilibrium of resources" was unquantifiable and unevaluable. Nonetheless, we shall see that this broad objective is reflected in many specific objectives of the plan audited by the field crew.

Wilderness. EISs on comprehensive land use plans naturally contain more objectives than the average EIS. One objective in the Grand Teton EIS clearly reflects the plan's broad "vignette of primitive America" objective. The master plan proposed 115,807 acres of designated wilderness. In 1978, after a detailed boundary survey, the wilderness proposal was adjusted to 122,604 acres. The Grand Teton wilderness was included on the 1978 official list of administration-endorsed additions to the national wilderness system and remained on that list. The Park Service has managed the area since 1976 exactly as it would have if it were officially designated wilderness. However, Congress never bothered to designate (or reject) this or any other National Park wilderness candidates because conflicts over proposed Forest Service and Bureau of Land Management (BLM) wilderness areas absorbed its attention.

This case provides an example of an outcome whose interpretation is definitionally complex. The outcome is factually clear: the current 122,604-acre area is essentially the same as the EIS's 115,807-acre proposal. The question is, is the area a "wilderness"? De facto, it is managed by the Park Service and used by the public as if it were a wilderness. But de jure, an official "wilderness" is an area statutorily designated by Congress. De facto, the forecast is accurate; de jure, it is inaccurate. As researchers familiar with federal wilderness policies, we classify this outcome as "complex, but basically accurate."

Residential Displacement. Another master plan objective involved a program of willing-seller, willing-buyer acquisition of private inholdings within the park and restoration of those inholdings to natural conditions. This objective is clearly an extension of the plan's general goal of preservationist management. Acquisition of park inholdings is often controversial (see Williams 1982) because many inholders bitterly resent losing their property, especially since they are just as enamoured of their lands' scenic qualities as is the Park Service. The EIS listed the number of inholdings as "in excess of 150, totalling 5,997 acres." The EIS writers noted, "Acquisition of the lands will result in a social impact upon the land owners; nearly 160 individuals are involved" (NPS 1975: 6,29). This forecast, alluding to an unspecified "social impact," is obviously vague.

Notwithstanding the aggressive tone of the EIS's discussion about "eliminating intrusive cultural features," the NPS proceeded very cautiously in acquiring inholdings and rarely displaced any bona fide residents. Interviews and NPS documents indicated that by 1983 the NPS had acquired 2,330 acres, or 39% of the inholdings at the time of the 1975 master plan. The major acquisition involved a complex 1983 land exchange with the park's largest inholder, Laurance Rockefeller, whose family has been a patron of Grand Teton. Rockefeller donated 1,221 acres to Princeton, Dartmouth, MIT, Vermont Law School, the Woodstock Foundation, and Sloan-Kettering Hospital, which then consumated a simultaneous land exchange and purchase arrangement in which Rocky Mountain Energy Company obtained 1,190 acres of BLM coal lands and NPS received the 1,221-acre park inholding.

Acquisitions from small-parcel inholders were invariably made on a willing-seller basis. Sales usually provide for a term occupancy or life estate allowing sellers to continue to live in their residences (NPS 1982: 30). From the creation of the park to 1983, only four parcels totalling 12 acres had been condemned. In short, because of the NPS's cautious approach, only a minority of the individual inholders in 1975 had been bought out, and because of the NPS practice of granting life estates, residents were rarely displaced. Thus, notwithstanding the vagueness in the EIS's allusion to "social impacts," the impact was classified as "less than forecast" because few of the 160 inholders had been displaced.

Recreation Visitation. Fostering outdoor recreation has been one of the two principal goals of the Park Service, but that use has always strained the Service's other goal, preserving parks' ecosystems. The Grand Teton plan emphasized the latter goal and thus proposed visitor restrictions such as limiting lodging and campground development to 1971 levels, barring autos and requiring bus transportation in parts of the park, and imposing back-country hiking and riding permits. The direct impacts of these controls involved three forecasts in the EIS. First, "these . . . restrictions would level off use" (forecast 6 in our coding). Second, the "restrictions would impose an inconvenience to the public" (forecast 5). Third, because of the restrictions, "visitor use and its attendant impacts will be shunted to areas outside the park. This will place pressures on the National Forests surrounding the park" (forecast 8). Annual recreation visitation to Grand Teton National Park and to the Bridger-Teton National Forest campgrounds near the park are graphed in Figure 6.9.

Annual visitation to Grand Teton averaged about 3 million people before the plan. In the three years after the FEIS, visitation increased to 3.9 million and then 4.2 million. Since a land use plan is implemented over a ten-year period, we are loath to weigh the 1975-1976 increase too heavily against the plan. Visitation leveled off to about 3.5 million during 1979-1982. As also shown in Figure 6.12, visits to national forest campgrounds near the park increased with increasing park visitation. If the phrase "will place pressures on the National Forests" means that national forest campground use will increase to absorb park campers while park visitation levels off, the visitation figures for the 1980s are roughly consistent with that implied forecast. Because we were unsure about the predictions implied by "level off" and "place pressures on," this impact was classified as "within the range of a vague forecast." We did not obtain adequate data about whether visitor restrictions imposed a significant inconvenience on the public, which is really the key impact in the three forecasts. Since people with different views about park preservation could disagree about the desirability of increased visitation, we rated the beneficiality of this impact "subjective."

181

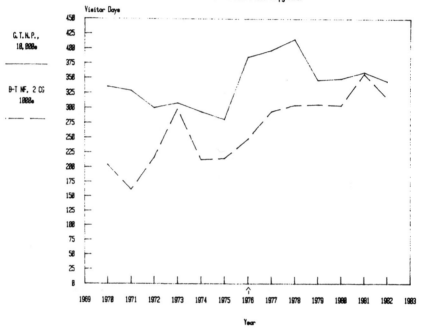

Grand Tetons: Recreation Visitors

Visitors to Park & N.F. Campgrounds

Figure 6.9

Grand Teton, Recreation Visitation: Recreation use of Grand
Teton National Park, western Wyoming, in 10,000 visitor-
days.   Comparison series is visitation at two adjacent
campgrounds on Bridger-Teton National Forest.   Forecast:
"These restrictions [holding overnight accommodations and
other facilities at 1971 levels] would level off use."
Grand Teton master plan approved March 1976 (arrow).   Impact
classified as "within range of vague forecast.

Sewage Treatment. Another objective listed in the master plan EIS was, "The Service plans to install sewage treatment facilities that will eliminate pollution problems now extant (although relatively minor) in the Jackson Lake area and at Moose [the park headquarters]." Interviews with biologists on the park staff confirmed that the sewage treatment facilities were indeed installed. Thus, treating this objective like a dichotomous, done-or-not-done mitigation measure, we classified the outcome as "close" to the forecast.

When asked about the effectiveness of the facilities, interviewees noted that the park had "spent a lot of money on sewage treatment," but that there were no real water-quality problems before the installation, and thus there are no current water-quality problems.[9] Ideally, physical monitoring data would have documented water quality before and after installation of the new sewage system. That is, reliable interview information may suffice to confirm that a facility was built, but interviewees' assertions are not good substitutes for physiographic data on the impact of the facility. Because of the uncertainty of the evidence about the second clause of the forecast ("will eliminate pollution problems"), this case received a secondary classification of "no clear impact, some impact predicted."

Forest Fire Management. The master plan's forest fire management section provides another example of the ecological philosophy underlying the Grand Teton plan. Revisionist forest fire researchers persuaded public lands managers during the 1970s that their traditional fire suppression policies ignored the natural role fire played in ecosystems before the twentieth century. Allowing natural fires to burn and setting "prescribed" fires would contribute, as the preface to the park's detailed fire management plan explicitly noted, to the restoration of a "vignette of primitive America."

The Grand Teton EIS contained four fire forecasts dealing with visibility, erosion, and wildfire impacts, plus the objective of carrying out a fire plan aimed at the "restoration of natural fire regimes." Interviews and NPS management documents confirmed that the new fire policy was implemented. The fire management plan divides the park into three zones, with all lightning-caused fires allowed to burn in zone I, some lightning fires allowed to burn in zone II, and all fires suppressed in zone III. Through 1981, only one of 22 zone I natural fires had been fought, and that was only partially suppresssed, and only half of the zone II

183

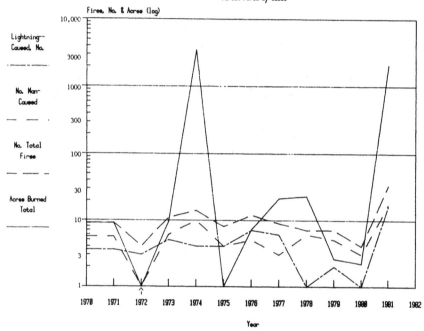

Grand Tetons: Fire Management

Forest Fires by Cause

Fires, No. & Acres (log)

Lightning-
Caused, No.

No. Man-
Caused

No. Total
Fires

Acres Burned
Total

10,000

3000

1000

300

100

30

10

3

1

1970   1971   1972   1973   1974   1975   1976   1977   1978   1979   1980   1981   1982

Year

Figure 6.10
Grand Teton, forest fires: Numbers and acreages of forest fires in Grand Teton National Park, western Wyoming, in logarithmic scale. Note: the 1970-1971 lines depict the averages during 1929-1971. Forecast: "[T]he most serious impact of [prescribed burning] is the possibility, however remote, of escape." Grand Teton master plan approved March 1976, incorporating fire plan first implemented in 1972 (arrow). Impact classified as has "not yet" occurred.

natural fires had been suppressed. According to the park superintendent, the new fire policy had become "well accepted." This outcome was thus classified as "close" to the EIS's objective.

The field crews also audited a forecast that read: "The most serious impact of using fire is the possibility, however remote, of escape. This possibility must be balanced against the very real threat of future uncontrollable holocausts if past fire management is continued, and unnatural fuel buildups are not dealt with" (NPS 1975: 31). This forecast is inherently difficult to audit because it is couched as a worst-case forecast, in which the impact is not expected to occur in any given year but might occur at some future time. The park's fire record, graphed in logarithmic scale in Figure 6.10, shows a significant increase in burning under the new policy. From 1929 to 1971, the park averaged 9.1 fires and 9.0 acres burned per year, with most fires caused by human carelessness. From 1974 through 1981, fires burned a total of 5,665 acres of the park, that is, 15 times the total acreage burned during 1929-1971. Two fires account for this increased burning: the Park Service let a 1974 fire burn 3,500 acres in zone I from July 19 to December 11, and let a 1981 zone I fire burn 2,000 acres from July 30 to October 14.

These figures simply indicate that the Park Service actually implemented its enunciated fire policy by letting two big backcountry fires burn for five and three months, respectively, the latter during 1981 when the park set a record for the most fires in a year. None of the 41 natural or 53 man-caused fires under the new plan escaped. However, since no one can know whether a fire may escape in the future, we are forced to classify this as a forecast impact that has "not yet" occurred. An alternate interpretation is that the second sentence of the forecast implied that an "uncontrollable holocaust" might occur if the old fire policy was continued. Since an uncontrollable fire had not broken out after 1973, the impact thus falls "within the range of a vague forecast." That interpretation neglects the fact that the 1974-1981 fires had not returned the whole park to natural fuel conditions. Nonetheless, the case received a secondary, "within the range of a vague forecast" classification. Since the effectiveness of prescribed burning and natural fire management is a matter of professional, scientific judgement, the beneciality of this case was rated "subjective."

Concessionaire Business. A major impact of the development ceiling imposed in the master plan is that businesses operating in the park would be disadvantaged relative to their competitors outside the park. Thus, the EIS forecast that the master plan's development ceiling "may have some highly localized adverse economic effects, primarily on park concessioners." That is, since lodge rooms, restaurants, and other services within the park would be frozen at 1971 levels, those concessioners' revenues would stagnate while their outside competitors would reap the advantages of visitation growth. Concessioners could not compensate for these limits by inflating prices, since the Park Service limits the reasonability of prices.

Concessioner revenue data and interview statements indicate, however, that the development ceiling did not have a particularly adverse effect on these businesses. As shown in Figure 6.11, lodging receipts remained relatively stable (discounting inflation), but concessioners' gross revenues increased substantially, roughly in step with increased park visitation during the period. According to a concessioner interviewee, the cap on the number of lodge rooms simply forced concessioners to adjust their operations by increasing their retail operations, directing their marketing more toward campers than lodge guests, booking conventions during nonpeak months, and so forth. On the whole, the adverse effect, if any, of the plan's development ceiling seemed limited, so this impact was classified as "less" than implied by the EIS's forecast.

Property Tax Base. Because of the general availability of tax data, property tax base forecasts were generally included on candidate audit lists. The Grand Teton tax base forecast involves a contingent impact, that is, an impact whose occurrence is contingent on another impact forecast in the EIS. In this case, the acquisition of the 160 parcels of inholdings (discussed above) was forecast to cause a "loss of land to the tax rolls."

The pattern of increases in Teton County's property tax base is perfectly normal, with property values outside Jackson increasing from $12.6 million in 1971 to $35.7 million in 1982. These tax base figures are not graphed because they are irrelevant to an evaluation of the forecast. First, the NPS has not yet acquired the bulk of the inholdings within the park. In particular, the largest acquistion, the Rockefeller ranch property, had been been donated by Rockefeller to tax-exempt institutions, so the 1983 NPS land exchange had no net effect on the tax base.

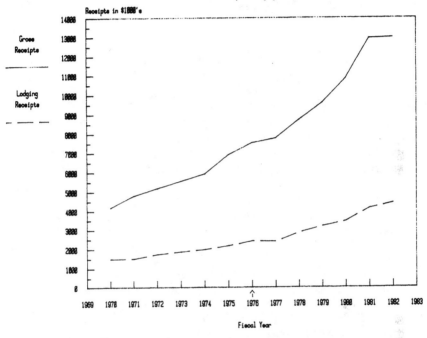

Grand Tetons: Effects on Concessioners

Figure 6.11
Grand Teton, concessioner economics: Gross receipts and
lodging receipts, concessionaire businesses in Grand Teton
National Park, western Wyoming.   Forecast: "[Holding]
further development of lodging, campgrounds, and certain
other service facilities . . . to present levels may have
some highly localized adverse economic effects, primarily on
park concessioners."   Grand Teton Master Plan approved
March 1976 (arrow).   Impact classified as "less" than
forecast.

Second, the tax impacts of any land acquisitions are dampened by two provisions. All federal land units provide payments in lieu of taxes (PILT) to local governments; Grand Teton's PILT payments rose from $22,084 to $28,841 during 1976-82. In addition, the park's organic legislation requires the federal government to continue paying property taxes at the full amount for ten years after acquisition and in evenly decreased amounts for the next 20 years thereafter. Since the contingent land acquisition had not been completed and the tax consequences of any acquisitions would be delayed for more than a decade, this case was classified as "no clear impact, though some impact forecast."

Cleveland Harbor Diked Disposal Site No. 12

This facility receives spoil material from ongoing maintenance dredging of the harbor at Cleveland, Ohio. Site no. 12 is located near the city's downtown lakefront and, when filled, was donated to the adjacent city-operated lakefront airport. The Corps' FEIS was filed in March 1973. Construction took place in 1974, and the site received dredging spoils from 1974 to 1979.

Cleveland disposal site no. 12 is a very routine project and, in two respects, turned out to be the worst EIS in our field sample. First, our coders identified only 12 forecasts in this EIS, the lowest number among the sample EISs, and none of these were quantified, so the EIS did not provide much choice in selecting seven candidate forecasts for the field audit.[10] Second, the field crews managed to collect little information on five of the seven impacts audited during the fieldwork. We have no quantified data, much less a time series, on any site no. 12 impact. Only one case's data quality is ranked "low," but the information on two other cases is paltry.[11]

In particular, we could not obtain satisfactory information on two key forecasts involving leakage of pollutants suspended in dredging sediments from the disposal site. This impact is most significant when dredging spoils come from a heavily polluted waterway, such as Cleveland's Cuyahoga River. The field crews were given conflicting information about a mitigation measure and related impact forecast on this issue. A local Corps official indicated that water-quality monitoring had been conducted and the raw data sent to the Corps' Buffalo District office. However, follow-up inquiries not only failed to secure these data, but met with the response that there was no record of such

Table 6.5
Summary: Cleveland harbor disposal site no. 12 forecast impacts in the field sample. Table includes both cases with adequate data and cases deleted because of inadequate data.

Category	I/M/O[a]	Impact Type	Primary Match[b]	Primary Data Type	Data Aptness
Biol.	Impact	Bird botulism	Range–VF	Verbal	Adeq.
	Mit.	Entrapment, fish	Close	Nominal	Low
Social	Impact	Municipal water system	NIm/NFor	Verbal	Adeq.
	Mit.	Public safety	Close	Nominal	Exact
Econ.	Impact	Powerplant operation	Underant–	Verbal	Adeq.
Subtotal = 5 impacts classified					
Phys.	Impact	Water quality, harbor	(no available data)		
	Mit.	Dike leakage monitoring	(inconclusive information)		
Total = 7 forecast impacts sought during fieldwork					

[a] Impact, mitigation, or objective.
[b] See Chapter 6 endnote 1, Table 7.1, or the text discussion of the particular impact for explanations of accuracy "match" codes.

monitoring. Thus, we did not know whether or not the mitigation had been done, nor did we have information other than unsubstantiated verbal assurances about the disposal site's impact on water quality.

Waterfowl Botulism. The site no. 12 EIS does not contain any compelling biological forecasts. The field crews thus audited an unclear discussion about possible infection of waterfowl by botulism. The EIS contains a lengthy, if evasive, response to an Ohio Environmental Protection Agency comment on this issue:

Botulism has long been a serious problem in the Great Lakes marshes where major kills number thousands of birds. Given the present state of knowledge, it is extremely difficult to prevent outbreaks of botulism in marshes frequented by waterfowl. . . . Once the organism invades an area, it remains in the sediments in

189

an inactive state until optimum conditions occur. . . .
If conditions in the diked disposal area are a causative
factor in outbreaks of botulism, interim management
measures will be taken. . . . Close cooperation will be
maintained, of course, with the appropriate agencies of
the State of Ohio (Corps 1973: 22).

This passage does not qualify as a full-fledged forecast,
and the corresponding passage in the EIS's impacts-of-the-
proposal section contains only vague sentences like "Use of
the diked area for nesting and breeding of birds is not
expected to be significant." The net effect of these
passages seems to be a vague forecast of "no impact."
   The field crews obtained information on this impact from
two sources. The local Corps engineer said he was not aware
of any outbreak of botulism in birds at any Cleveland Harbor
disposal site. The City of Cleveland's Commissioner for
Environmental Health Services confirmed that there was no
record of any bird botulism cases in the harbor. This
information is certainly scanty (though, since it was based
on multiple sources, the information was rated "adequate"
under our general criteria). The impact was classified as
"within the range of a vague forecast," with a secondary
classification of "no clear impact, 'no impact' forecast."

   Fish Entrapment. The site no. 12 EIS proposed a
routine biological mitigation measure: "The real problem of
fish entrapment has been recognized and the State of Ohio's
Department of Natural Resources has, by written letter,
agreed to cooperate with the Corps, including seining as
necessary, to preclude any large fish kill" (Corps 1973:
14). The field crew's interview with a local Corps official
confirmed that the Corps had seined (i.e., screened fish
with a net away from the dredging intake) as a routine part
of its dredging operations. Despite its vagueness about the
nature of the DNR-Corps cooperation, this forecast was
classified as "close" to the actual (if trivial) mitigation.

   Municipal Water System. The Corps (1973: 14) also
noted that a municipal water tunnel is located under
disposal site no. 12, but forecast, "The new facility will
not significantly affect the existing loading on the
tunnel." This case also illustrates our desperate search
for auditable forecasts in this EIS.
   Two interviewees confirmed that the disposal facility
had not affected the water intake tunnel, and the case was
thus classified "no clear impact, 'no impact' forecast."

190

Since the disposal facility merely displaced the weight of the site's water with a comparable amount of sediment, subsidence damage would have been unlikely. Thus, the case received a beneficiality rating of "trivial" and a secondary classification of "intuitively obvious accuracy."

Public Safety. The Corps (1973: 15-16) noted that it would take such safety precautions as installing warning signs and fences, concluding "The inaccessibility of the structure by the general public reduces any danger to the public. . . ." This mitigation is, of course, as routine as the fish seining measures covered above. A site tour and interviews with a local Corps official and the Cleveland harbor police confirmed that the site was indeed fenced and marked with warning signs. Occasionally duck hunters used the site as a duck blind before being chased off by the harbor police. Police records confirmed that there had been no accidents or safety incidents on the site, except for an airplane that crashed into the site during an airshow.

Power Plant Operation. The most interesting impact of disposal site no. 12 was essentially ignored by the Corps' EIS writers. The midwest office of the U.S. Environmental Protection Agency noted, in its comments on the site no. 12 DEIS (Corps 1973: appendix), "There is no discussion relating to the possible adverse conditions resulting from reduced circulation of water currents east of the structure. A power plant uses this harbor area as a source of cooling water and also discharges it back into the same general area." The Corps (1973: 25) denied this impact would occur: "It is expected that circulation of water, induced by changes in water levels of the river and harbor, will be substantially the same, before and after construction. Therefore, no adverse conditions are anticipated." (The Corps' responses to EPA's comment letter were generally quite snide. For example, EPA's commenter pointed out valid concerns about pollution containment within the site's dike and recommended a periodic monitoring program to ensure water quality standards were not violated. The Corps' response was, "We concur and would recommend that EPA do the monitoring, for objectivity.") Because this interchange was buried deep in the FEIS's comments-and-responses section, the coders had passed over it during the content analysis.

EPA's commenter turned out to be correct. Disposal site no. 12 is immediately offshore from the cooling water intake ports of the Cleveland municipal power plant. According to an interviewee in the city power department, site no. 12

affects water circulation so that the plant's intake water is heated by the plant's thermal discharge, making the intake water too warm for effective cooling. The Cleveland municipal power plant has been closed for several years because of economic problems, controversies over city management of the plant (which were regarded as symptomatic of Cleveland politics during the tenure of Mayor Dennis Kucinich), and other problems. However, site no. 12's circulation effects would pose significant engineering problems in reopening the plant. Because the Corps had ignored this issue, which reflected an already well-known understanding of coastal processes,[12] the case was classified as being in the "wrong direction," "underanticipated," and "more adverse than forecast." This forecast is one of the worst among the 239 cases in the field sample.

ADDITIONAL REPRESENTATIVE IMPACTS

The Paint Creek, Sequoyah, Grand Teton, and Cleveland harbor projects provide examples of all but four of this study's primary accuracy classifications, and three of the classifications not represented among these projects contain only one or two cases throughout the whole field sample. The following set of examples complements the cases discussed above. This section covers all but four of the primary forecast-impact classifications. The projects covered above exhausted the examples of the "underanticipated-beneficial" and "underanticipated-adverse" classifications. The third classification, "impact has not yet occurred," did not seem sufficiently complicated to merit an additional example beyond the Grand Teton forest fire case. Finally, both examples of the fourth code, "complex but apparently inaccurate," seemed too uninteresting to cover in detail.[13] The cases presented below present either representative or particularly striking examples of forecast accuracy classifications.

Accurate Forecasts

The most accurate classification in the audit scheme, "forecast close to actual impact," is well represented in the four projects' cases reviewed above, so we present only two "close"-rated cases below. In addition, we present two forecasts to demonstrate that some cases that did not receive a "close" rating were still as fairly accurate.

Monticello Nuclear Station: Property Taxes. Construction of the Monticello station began before the passage of NEPA, and the plant received its full-power operating license in January 1971. The _ex post facto_ FEIS on the plant, while enumerating the economic benefits of the plant, noted that Northern States Power would provide "financial support of nearby communities through annual tax payments of about $2,300,000/yr" (AEC 1972: XI-15). This case involves a simple quantified forecast that proved to be accurate.

The actual property taxes paid by the company on the Monticello plant are graphed in Figure 6.12. The company paid the forecast amount of taxes in the years immediately following the EIS, with plant taxes climbing over $3 million in step with the property value inflation of the late 1970s. This forecast was thus classified as "close" to the actual impact. The plant's taxes had a visible effect on the public facilities of the small town of Monticello; in particular, the school system was able to construct a large, beautiful campus. Thus, the impact was classified "as beneficial" as implied by the forecast. As noted regarding the comparable Sequoyah plant forecasts discusssed above, however, projecting near-term property tax payments for a completed plant is extremely easy.

Weymouth-Fore and Town Rivers: Rock Blasting. The Weymouth-Fore and Town rivers share a channel to Massachusetts Bay. A 1974 Corps EIS proposed to remove rock outcroppings to complete channel deepening. The safety of blasting operations was the most controversial social impact of the project. The key safety forecast involved blast vibration: "The response of oil tanks to vibration caused by blasting is therefore more critical than for buildings and accordingly the maximum specified permissible particle velocity at the oil tanks will be 0.5 inches per second. Blasts will be recorded by seismographs and reports will be submitted daily" (Corps 1974a: 3-6).

The Corps' consultant conducted seismic tests during each blast, recording vibrations at the nearest onshore building (which was usually closer to the blast than were the oil tanks). Blast vibrations averaged 0.12 inches/second particle movement velocity during 36 blasting days from October 1974 to July 1976, and only three readings exceeded 0.23 inches/second. One reading equaled 0.53 inches/second, but rounded down to the forecast 0.5, and thus did not violate the EIS limit. For comparision, during the first monitoring, the contractor measured the vibration caused by a small child jumping on a carpeted floor at the same house

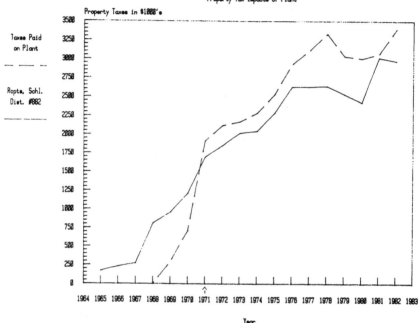

Monticello Nuclear Generating Plant
Property Tax Impacts of Plant

Figure 6.12

Monticello Station, property taxes: Total property taxes paid on Monticello Nuclear Station (dashed line), plus total tax receipts by Monticello Independent School District 882 (solid line). Forecast: "Benefits [of plant include] . . . financial support of nearby communities through annual tax payments of about $2,300,000/yr." Plant received full-power operating license January 1971 (arrow); no clear change in plant's license status following 1972 FEIS in sample. Impact classified as "close."

194

as 0.52 inches/second.

While the actual impact of Weymouth-Fore rock blasting was insignificant, the case illustrates several complexities encountered in auditing forecast accuracy. First, the seismic readings are excellent time-ordered—but not interrupted time-series—construction-phase data. Second, like many cases in the sample, the forecast is couched in terms of a limit. Since all readings are less than or equal to the limit, the impact is rated as "close" to the forecast. However, since actual blast vibration was almost always less than half the 0.5 limit, the impact is also rated as "less adverse" than the forecast implied. Third, this case involves both an impact, the vibration effects of blasting, and a mitigation measure, the monitoring of blast vibration. Thus, this case receives a secondary match code of "mitigation done."

Jackson Airport: "Nonconforming Use." As described in Chapter 3, the proposed enlargement of the Jackson airport was one of two very controversial projects in the field sample. The real issue for most protagonists in this controversy was whether an airport should be located within national park boundaries. The airport was built in 1941 on BLM and private land outside the then-existing national park. In 1943 and 1950 the airport site was included in a two-step expansion of the park. The park's legislation authorizes the Department of the Interior to permit the airport, and the Park Service granted the airport a 20-year special use permit in 1955. However, the existence of an airport clashed with Park Service policies of the 1970s (e.g., the "vignette of primitive America" objective of the contemporaneous Grand Teton master plan). The Chamber of Commerce vigorously supported the airport as a key gateway for its tourist trade, but environmental groups, particularly the Sierra Club, demanded the removal of the airport.

The Park Service's 1974 FEIS's rejection of jet service should be recognized as an implicit attempt to induce the airport to move elsewhere. The forecasts that hint at the agency's real objective include an observation about jet use: "Retention of the runway length of 6,305 feet will restrict regularly scheduled commercial use of the airport by the Boeing 737. [The B-737] could not be operated economically from the present Jackson Hole Airport. . . ." (NPS 1974: 70, emphasis added); and a comment on aesthetics, "Construction of additional airport facilities would result in increased visual intrusions" (NPS 1974: 79).

Federal decisionmaking about Jackson airport has escalated frequently to high levels of the executive branch. Improvements in aircraft technology made jet service to Jackson feasible without a runway extension. Following a separate Department of Transportation EIS, a spate of Sierra Club suits (cf. Ruckel 1980, 1985), a referral of the dispute to the executive office of the president (CEQ 1985: 528), and other interagency political maneuvers, jet service into Jackson eventually began in June 1981. As an important part of the interagency politics, Secretary of the Interior Andrus revised the airport's permit in August 1979, ruling that the airport was a "nonconforming use" within the park and issuing the airport a new permit terminating in 1995. However, the Reagan administration's Secretary of the Interior, Jim Watt, reversed Andrus's decision in November 1982 and awarded Jackson airport a new, liberal lease through the year 2033.

Clearly, this highly politicized case poses evaluation problems. The EIS does not contain a clear forecast about the real objective of the proposal, and the 1974 FEIS represents only one stage in a convoluted sequence of decisions. In our judgment, Secretary Andrus's 1979 decision implemented the spirit of the Park Service's professional stance in the Jackson FEIS. Thus, notwithstanding Secretary Watt's subsequent reversal, we classified the accuracy the Park Service's implied forecast about its objectives as "complex, but arguably accurate." Needless to say, we also coded the beneficiality of this stream of decisions as "subjective."

U.S. 53: Traffic Safety. This study gives its best rating to cases in which a forecast closely matches an actual impact. Readers should understand that some very good outcomes are classified as relatively inaccurate forecasts. The traffic safety objective of a highway project in the sample is one such misleading primary match classification.

U.S. 53 is the major highway from Duluth south to west-central Wisconsin. The FEIS in the sample described a primary objective of the upgrading of the Barron County segment of U.S. 53 to a four-lane highway as follows: "The upgrading of USH 53 to freeway standards should reduce the number of accidents by approximately one-half, . . . with the number of fatal accidents reduced by one-third to one-half" (Wisconsin DOT 1975: 66). Work on the U.S. 53 project recommenced in May 1976, and was completed in November 1977.

196

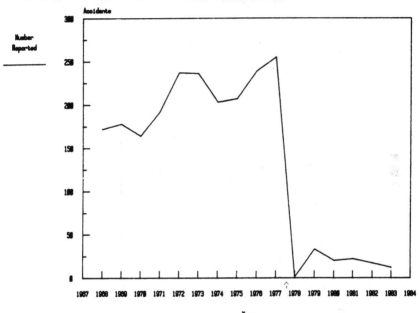

U. S. Highway 53: Traffic Safety
Number of U.S. 53 Accidents, Barron Co.

Figure 6.13
U.S. 53 and U.S. 8 traffic safety: Accident rate, U.S. 53, Barron County, northern Wisconsin. Forecast: "The upgrading of US53 . . . should reduce the number of accidents by approximately one-half . . ., with the number of fatal accidents reduced by one-third to one-half." U.S. 53 segment completed November 1977 (arrow). Outcome classified as "exceeds" forecast.

As shown in Figure 6.13, the accident rate on the project segment of U.S. 53 dropped precipitously from over 200 accidents/year before the upgrading to 17.5 accidents/year afterwards. The county sheriff and state highway department interviewees confirmed that the project had transformed this highway from a "deathtrap" into a safe road. The project thus proved to be very effective, and the general tenor of the forecast, predicting a significantly safer road, was accurate. However, since the improvement in safety was much greater than the forecast "one-half," this case was classified as "impact exceeds forecast." The code that rewards the Wisconsin highway planners for their effective project is the case's beneficiality rating that the outcome was "more beneficial than predicted."

Vague or Unusual Forecasts' Accuracy

It is often more difficult to interpret some of the vague forecasts in the study's sample than to determine the postproject impact. One example that follows represents the typical problem posed by a vague forecast. A second case covers the obverse situation, in which the forecast is not mysteriously vague, but intuitively obvious.

Shepard Park Development: School Enrollment. The Shepard Park development consists of eight blocks of high-rise or mid-rise residential buildings in an already urbanized section of St. Paul, Minnesota. One social forecast in the 1977 Department of Housing and Urban Development's Shepard Park EIS dealt with the impact of this 700-unit development on the school system: "The St. Paul Public Schools have been undergoing a period of rapidly declining enrollments. . . . [Thus,] the School District has sufficient capacity to handle the additional students from the Shepard Park development" (HUD 1977c: 68). Figure 6.14 shows enrollments for both the local school, Homecraft, and the St. Paul school system.

School enrollments indeed declined both before and after the Shepard Park EIS, with an even steeper decline at Homecraft School than in St. Paul as a whole. Homecraft's enrollments leveled off in 1978; after falling from 585 pupils in 1971 to 317 in 1977, enrollment declined gradually to 264 in 1983. However, this effect cannot be clearly attributed to Shepard Park: enrollment stabilized during fall of 1978, but the first Shepard buildings were occupied later and contained either senior citizens' units with no

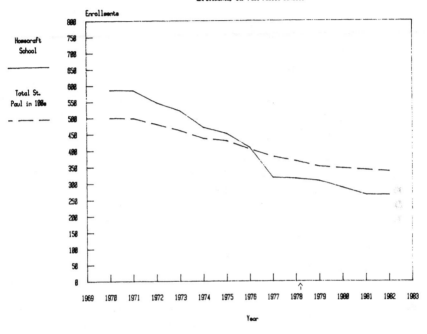

Shepard Park: School Enrollments

Enrollments, St. Paul Public Schools

Figure 6.14
School enrollments, Shepard Park apartment building complex, St. Paul, Minnesota. Forecast: "The School District has sufficient capacity to handle the additional students from the Shepard Park development [because of] . . . a period of rapidly declining enrollments." Figure plots official (late September) enrollments at Homecraft School (solid line), within whose boundaries the development is located, and St. Paul school system total enrollment (dashed line). Major buildings in the Shepard Park development were completed and occupied at about one-year intervals from December 1978 (arrow) to August 1982. Impact classified as "within the range of vague forecast."

school-age residents (completed November 1979) or "luxury" apartments or condominiums with relatively few school-age residents (three buildings, completed December 1978, November 1980, and August 1982). In any case, how could one evaluate any data in terms of the forecast? The phrase "has the capacity" allows any Shepard-caused impact ranging from a net decrease in enrollment to a doubling of enrollment from the 300-pupil level of 1977-1979 back to the 585-pupil level of 1970-1971. Solely by virtue of its lattitude, the forecast is not inaccurate.

Illinois Beach State Park: Habitat Preservation. As noted in Chapter 3, about 5% of the forecasts in the sample EISs deal with incidental impacts--that is, consequences that must logically occur immediately incident upon the execution of the EIS's proposed action. The field teams did not generally seek to audit such incidental impacts, but occasionally accepted such forecasts for fieldwork because no other satisfactory candidates fit one of the impact categories specified by our evaluation protocol. The Illinois Beach State Park expansion provides one such case.

The 1973 Bureau of Outdoor Recreation FEIS in the sample covered a grant to the Illinois Department of Conservation to double the size of Illinois Beach State Park, on Lake Michigan in the northeastern corner of the state. The principal objective of the land acquisition, as repeated in several EIS passages, was to preserve wetlands and wildlife habitat from residential development. Most parcels in the new unit of the park were acquired from 1972 to 1979. Once the parcels were acquired--baring any change in park management that would contravene the proposal described in 1973-- the wetlands and wildlife habitat were preserved. A site visit confirmed that the wetlands and habitat are still preserved. This forecast thus received a classification of "intuitively obvious accuracy" because the impact occurred immediately on implementation of the proposed action. A wetlands professional in the region indicated that the wetlands and habitat are fairly significant, so the outcome was also classified "as beneficial as forecast."

The "No Clear Impact" Classifications

The key criterion for assigning a "no clear impact" classification is that an impact is not clearly associated with the cause specified in the EIS's forecast. Two classifications are based on a "no clear impact" coding of the actual impact; one applies to cases in which the EIS had forecast no significant impact, and the other to cases in which the EIS had forecast some change. Even though the available data do not permit firm conclusions about these cases, the former classification naturally appears to be more accurate than the latter.

Monticello Station: Ambient Radiation. Radiation exposure is considered the primary impact of a nuclear facility, and nuclear plant EISs generally contain several radiation forecasts. The Monticello EIS, for example, contained eight radiation forecasts, and the field teams audited three radiation impacts. A set of three forecasts dealt with human exposure to ambient atmospheric radiation. These forecasts were spread over six pages and were quantified in great detail. They concluded, "The integrated total-body dose to the population living within 50 miles of the plant from submersion in radioactive gaseous effluents was estimated to be about 1.5 man-rem/yr with the augmented system in operation" (AEC 1972: V-31), which the EIS then contrasted with the natural background radiation dose of 150 rems. In other words, this forecast implies the plant would have no real impact on ambient radiation levels.

Monitoring data on radiation emissions and ambient conditions, which the NRC requires from nuclear licensees, provide more valid indicators for auditing this forecast than would data on man-rem doses. First, a separate forecast in the Monticello EIS stated that the plant's "augmented system" would decrease radioactive emission rates by more than an order of magnitude. Emissions decreased from an average of 8 x 10^5 curies during 1971-1974 to 6 x 10^3 after 1975, so this contingent forecast was accurate.

Second, as shown in Figure 6.15, the quarterly averages of daily ambient radiation readings are generally consistent with the the EIS's forecast. Gross beta radiation averages after December 1970, when the Monticello reactor went critical, are generally below or within the range of 1968-1970 preproject variance. Monitoring data could not be located for 1974, when the plant experienced its worst radiation emissions problems (i.e., 1.6 x 10^6 curies). In addition, as with all nuclear monitoring reports reviewed during the fieldwork, the plant's consultant regularly noted that Chinese atmospheric bomb tests caused periodic sharp increases in ambient gross beta readings. Thus, the EIS's forecast that the plant would not perceptably increase ambient radiation levels was classified as "no clear impact, 'no impact' forecast."

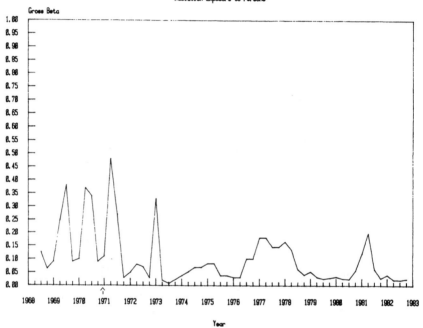

Monticello Nuclear Generating Plant
Radiation Exposure to Persons

Figure 6.15
Monticello Station, Ambient radiation levels. Forecast: "The integrated total body dose to the population§.§.§. was estimated to be about 1.5 man-rem/year," contrasted with a natural background dose of 150 rems. Figure plots quarterly average gross beta radiation in picocuries/m³, 1968-1980. Reactor critical December 1970 (arrow); final EIS filed November 1972. Case classified as "no clear impact, 'no impact' forecast."

Jackson Airport: Passenger Loads. The Park Service's Jackson airport EIS discussed allowing jet service by extending the airport runway, constructing a new control tower, and installing other technical improvements. Following the 1974 FEIS, the Department of the Interior deferred the runway extension, new tower, and jet service. The other technical improvements were installed during 1975-1976. Improvements in jet technology made jet service feasible without a runway extension. Thus, after another round of controversy (discussed above), the FAA approved B-737 jet service on a trial basis. The first scheduled jet flight into Jackson landed in June 1981.

The EIS indicated that the main objective of allowing jet service was to improve airline efficiency. This objective was couched as a model quantified forecast:

> The tendency for the total number of flights to decrease due to increased capacity [of jets] will likely be offset to some degree by an increase in the total number of passengers. . . . Speas Associates estimates that jet service would result in a forecast of 57,000 enplanements by 1985; this estimate is about 24% greater than the forecast of enplanements without jet service capabilities (NPS 1974: 102).

As shown in Figure 6.16, the actual passenger loads through Jackson airport do not exactly conform to this forecast. Jackson's passenger loads increased steadily from 1973 to 1985, eventually reaching 65,613 enplanements in 1985. However, this increased passenger load is not clearly associated with the onset of jet service per se. First, the initial sharp passenger increase came in 1978-1979, a year after the installation of an instrument landing system and other technical equipment. The 1985 passenger level of enplanements approximately equals a linear extension of the trend associated with the post-1976 increase. Second, passenger loads decrease during 1981-1982, immediately after the beginning of jet service. This decrease seems to be a spurious result of instability in the industry following the deregulation of civil aviation. Jackson's routes were constantly reassigned to Frontier, Western, Transwestern, and other carriers during the period. Airline passenger loads nationally also declined sharply during 1979-1981 in a pattern very similar to that in Figure 6.16.[14] Third, the 1983-1984 increase in enplanements occurred two years after jet service began. This increase is in line with the 34% increase in commercial passengers nationally following

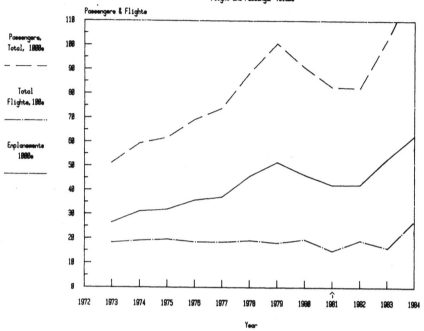

Jackson Airport: Air Traffic

Flight and Passenger Volume

Figure 6.16

Jackson airport, airline passenger volume: Enplanements
(solid line), flights, and total passengers, Jackson
airport, Teton County, western Wyoming. Forecast: "The
tendency for the total number of flights to decrease due to
[jets'] increased capacity will likely be offset . . . by an
increase in the total number of passengers. . . . Jet
service would result in a forecast of 57,000 enplanements by
1985; this estimate is about 24% greater than the forecast
of enplanements without jet service capabilities." New
equipment, notably instrument landing system, operational in
1976; scheduled jet service began June 1981 (arrow).
Classified as "no clear impact, though some impact had been
forecast."

deregulation (Chapman 1986).

A separate forecast indicates that the Jackson ski industry was expected to be the prime beneficiary of certain airport improvements, primarily the instrument landing system, but not jet service per se. After a poor snow year in 1976-1977, monthly ski-season passenger counts increased significantly beginning in December 1977 and continued through the 1980-1981 season. The data on ski area use suggest this increase contributed to annual increases of about 11,000 skier visitor-days, though the year-to-year variance in ski visits is affected by snow conditions and systematic shifts among ski areas in the Rockies.

In other words, passenger loads increased, and indeed exceeded the 1985 forecast level by 15%. But those increases are not clearly associated with jet service, as specified by the forecast. We suspect that the increase in Jackson's passenger loads is largely accounted for by improvements approved by the Park Service and installed in 1976, particularly the instrument landing system, plus the local share in the 56% nationwide increase in passenger volume during 1976-1984. Thus, this case was classified as "no clear impact, though some change forecast." Jet service into Jackson, as distinct from the 1976 instrument systems, was also rated as "less beneficial" than forecast.

The "Exceeds" and "Less" Classifications

Certain impacts differ significantly from their forecasts, but are not so inaccurate as to receive our lowest accuracy classification. These impacts are in the predicted (i.e., "correct") direction but with a magnitude of change noticeably greater than forecast. The criterion for coding such impacts relies simply on numerical values: if the number measuring the impact is greater than the forecast number, the impact is rated as "exceeds the forecast"; if the number is less, the impact is rated "less than forecast." Thus, there is no conceptual difference in forecast accuracy between "exceeds" and "less" classifications.

Monticello Station: Generating Reserves. The fundamental justification for new power plants is that the plant's capacity is needed to meet load growth and maintain a safe generating reserve. The Monticello EIS's objective is typical of this argument:

[The] generating capacity of the Monticello plant

comprises practically all of Northern States Power's generating reserves in the 1972-1976 period. By 1973 the reserve criterion of 12% will not in fact be met by the generating capacity of Monticello, and projected yearly increases in reserve deficiencies will necessitate that Northern States Power bring into being new generating capacity. . . . [T]he projected peak load reserves of the Upper Mississippi Power Pool are seen to be 11.5 to 13.3 percent in the period 1972-1976. Without the Monticello Plant, this reserve would be reduced to between 3.5 and 5.8 percent, far below the reserve required for dependable service, and incurring the likelihood of inability to meet load demands during peak periods (AEC 1972: XI-1 - XI-2).

As shown in Figure 6.17, instead of the forecast's target reserve of 12%, the actual Northern States Power reserve ballooned in the late 1970s into the 20-40% range, averaging 29% during 1975-1982. Overinstalled generating capacity was a problem for most utilities during this period, when electrical load growth decreased as a result of energy conservation and other economic changes in the mid-1970s. The Monticello case is simply typical of a national trend in the utility industry.

This case, however, poses minor accuracy coding problems. First, the direction of the actual outcome is ambiguous because the forecast involves a single number target. That is, is a 29% reserve (or a 2% reserve) in the "correct" or "wrong" direction, relative to a 12% target? Missing the target on either the high or low side involves trade-offs among planning objectives, such as minimizing capital costs to ratepayers versus minimizing the probability of service disruption. We indicate such ambiguous cases with a code of a "bad" direction of the impact. Second, a 29% reserve is numerically greater than 12%, so this case was classified as "impact exceeds forecast." (However, if the indicator had been actual electrical generation as a percent of capacity, the case would have received the equally inaccurate "less than forecast" rating.) Finally, despite pedantic concerns about numerical directions, we confidently code this outcome as "less beneficial" than forecast because the actual reserve so exceeds the public utility standard of 12%. Indeed, critics of overinstalled nuclear capacity might argue this forecast was so wrong that it merits a worse accuracy code. This case did not receive an "inconsistent" classification, however, because the EIS's forecast stressed the criterion of

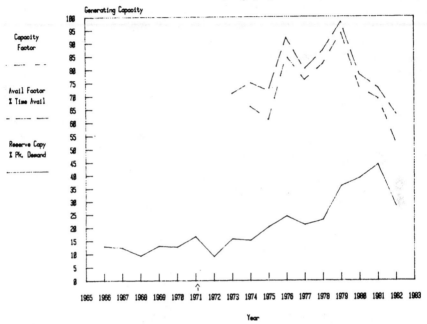

Figure 6.17
Monticello Station, reserve capacity: Electrical generation
reserve capacity, Northern States Power Company, in
percents. Forecast: "The projected peak load reserves of
the Upper Mississippi Power Pool are . . . 11.5 to 13.3
percent in the period 1972-1976. Without the Monticello
Plant, this reserve would be reduced to between 3.5 and 5.8
percent, far below the reserve required for dependable
service." Plant received full-power operating license
January 1971 (arrow); no clear change in plant's license
status followed 1972 FEIS in sample. Impact classified as
"exceeds" forecast.

reliable service and the Monticello plant certainly allowed Northern States Power to meet load demands.

Illinois Beach: Tax Base. The Illinois Department of Conservation (IDoC) acquired 1,269 properties in its expansion of Illinois Beach State Park. It acquired 266 residential parcels (50 of which were valuable lakeshore properties threatened by erosion), mostly between 1972 and 1975, and 1,003 vacant parcels, mostly between 1973 and 1979. In one of two interrelated tax forecasts, the Bureau of Outdoor Recreation (BOR 1973: 22) noted that, because of the acquisition, "approximately $150,000 per year will be lost to the community initially in taxes"; the EIS then broke this tax loss down among Winthrop Harbor (the local municipality), Benton Township school districts, and other taxing bodies within Lake County.

As shown in Figure 6.18, the short-term tax revenue loss caused by the Illinois Beach expansion was more severe than forecast. The EIS forecast the Zion-Benton High School would lose $59,000, but the high school tax receipts from Benton Township actually fell by $136,818 from 1971 to 1975. During the same period, tax revenue from the affected Winthrop Harbor Grade School districts dropped by $110,222, rather than the $34,000 decline forecast in the EIS. The Village of Winthrop Harbor's tax receipts remained relatively stable during the acquisition, increasing from $94,033 in 1971 to $107,679 in 1975. However, as shown in Figure 6.19, Winthrop Harbor actually lost $3.8 million in tax base during the Illinois Beach acquisition, and made up for the tax base loss by increasing tax rates.

Several control series confirm that the park expansion significantly affected Winthrop Harbor's tax base. As shown in Figure 6.19, at the same time that Winthrop Harbor lost 25% of its tax base, Lake County assessments increased normally. The City of Zion, immediately south of Winthrop Harbor (plus the Zion Township part of the Zion-Benton High School District), experienced a property tax windfall during the period because Commonwealth Edison Company opened a new nuclear power plant there.

The EIS also argued that the short-term loss of property tax base would be compensated by long-run tax gains:

> [T]he immediate impact would be offset within five years by added income received from people using the recreation area. [IDoC] anticipates an additional visitation of 1,000,000 persons annually once initial development occurs. If each spends $1.50 per visitor

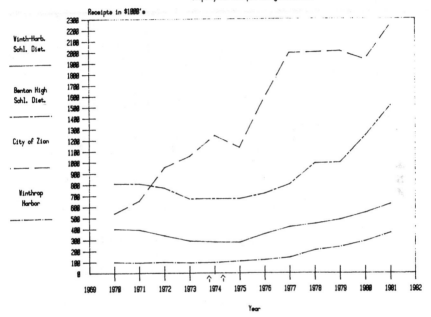

Illinois Beach: Local Tax Base-Receipts
Property Tax Receipts-Villages & Schools

Figure 6.18
Illinois Beach expansion, short-term tax base loss:
Property tax receipts, four taxing districts, Lake County,
Illinois. Forecast: "The resulting annual loss of tax
revenue would include $59,000 to the Zion-Benton High School
District; $34,000 to the Winthrop Harbor Grade School
District; . . . $16,000 to Lake County; . . . and $10,000 to
the Village of Winthrop Harbor." Zion tax revenues were not
affected by the expansion, and are plotted for comparison
purposes. Most improved properties acquired 1972-1975
(arrows); most unimproved properties acquired or condemned
1973-1979. Impact classified as "exceeds" forecast.

day, . . . this would bring an additional $1.5 million annually to the regional economy. . . . This would be reflected in [higher sales] tax revenues. . . . Also it can be expected that land values adjacent to the park will increase, thus increasing [property] tax revenue (BOR 1973: 23).

As shown in Figure 6.19 Winthrop Harbor's tax base had not recovered by the time of our fieldwork. After the major period of acquisition ended, Winthrop Harbor's assessments increased more slowly than did property values in Lake County as a whole. Local tax revenues did not recover within the forecast five years because the intervening contingencies in the forecast did not materialize. The EIS stated that the new unit would be developed by 1974-1975 and would include swimming, hiking, camping, and picnic facilities, plus a marina. By the time of the fieldwork, only a small picnic area had been built. Indeed, the master plan for the expanded park was only released in April 1982, and that plan was not brought to fruition until 1984-1985. Construction of the marina, the key development, did not begin until October 1986. Illinois Beach visitation thus only increased from 1.27 million in 1970 to 1.58 million by 1982, sales tax revenues had not blossomed, adjacent property values had not increased, and Winthrop Harbor had not recouped its tax losses. As a Zion-Benton Chamber-of-Commerce officer charged in a field interview, the acquisition placed a major burden on Winthrop Harbor taxpayers, but the Department of Conservation had not delivered on its promised offsetting recreational developments.

The Illinois Beach EIS's tax forecasts received nominally different classifications: the short-term property tax case was coded "impact exceeds forecast" and the long-term tax revenue "impact less than forecast." The net effect in each case was the same, however. The former was coded "impact more adverse than forecast" and the latter "impact less beneficial than forecast."

Indeterminant Accuracy Classifications

Several cases could not be conclusively audited because unusual circumstances obscured the impact. These circumstances do not include inadequate data, which caused 100 cases to be dropped from the audit. (See Chapter 5.) Indeed, we often obtained extensive data or interview information on these cases. Two Hiwassee unit plan cases

210

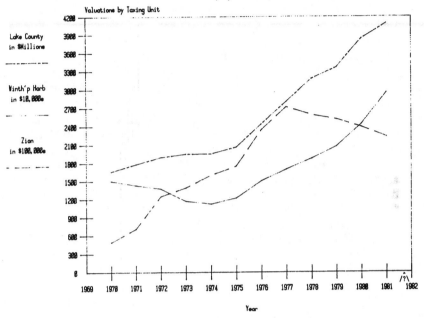

Illinois Beach: Local Tax Base
Property Assessed Valuations

Figure 6.19
Illinois Beach expansion: Long-range tax base enhancement.
Forecast: "After an initial period of adjustment, there is
reason to believe the local tax base, the economy, and the
community will expand rather than decrease as a result of
park expansion." Figure plots assessed valuations of
property in Winthrop Harbor (solid line), the jurisdiction
of the expansion, and, for comparison, Zion (dash line) and
Lake County, Illinois (dash-dot line). Master plan for
expansion unit not drafted until 1982, after the data series
ended; major expansion-unit recreation facilities not
constructed in 1984. Impact classified as "less" than
forecast.

illustrate circumstances that cannot support a firm conclusion about the impact.

"Spurious": Hiwassee Unit Plan, PILT. Inaccurate forecasts can often be attributed to EIS writers wrongly assuming that "everything else would be equal." In certain cases factors unrelated to the project were so blatant that we assigned these cases a separate classification. The Hiwassee unit plan covered a planning unit comprising about half a ranger district on the Cherokee National Forest in southeastern Tennessee. The plan proposed a slight reduction, from 4.8 to 4.5 million board feet annually, in the unit's programmed timber sales. Since federal land is not subject to local property taxes, the agency returns 25% of timber sale receipts to local governments as payments in lieu of taxes (PILT). The EIS thus noted, "This reduction in receipts will in turn reduce the amount of money that the county will receive from the Forest Service as a percentage of timber sale receipts" (Forest Service 1975: 19).

The plan was approved soon after the 1975 filing of the plan's FEIS, and was implemented through the subsequent ten-year plan cycle. The Hiwassee's PILT actually increased from around $24,000 to over $50,000 the year after the plan was implemented. The 1976-1977 increase had nothing to do with changes in Hiwassee timber sales (Figure 6.20, solid line). Timber sales fluctuated widely between $51,000 and $297,000, with year-to-year shifts averaging $88,000 between 1968 and 1982.

Rather, the levels of PILT payments were purely an artifact of federal procedures for calculating PILT payments. Timber sales on the Hiwassee unit had little or no influence on payments to the county in which the Hiwassee unit is located because all eligible receipts on the Cherokee National Forest were averaged together and each county received payments proportional to its percentage of national forest land. Moreover, Congress changed the formula for calculating PILT payments effective with fiscal year 1977. Several arcane changes (including timber purchasers' road-building credits, reforestation funds, and uncut timber credits in the receipts base) had the effect of doubling the PILT base. Therefore, this case's inaccuracy was classified as "spurious" since there was absolutely no connection between the actual outcome and the land use plan or EIS forecast.

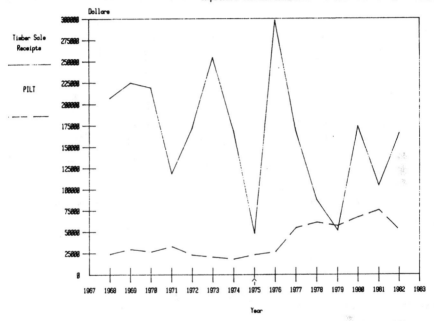

Hiwassee: Timber Production and PILT

Comparison of PILT with Timber Sales

Figure 6.20
Hiwassee unit plan: Federal payments in lieu of taxes to
Polk County, Tennessee, for lands in Cherokee National
Forest lands. Forecast: "This reduction in receipts will
in turn reduce the amount of money that the county will
receive from the Forest Service as a percentage of the
timber sale receipts." Figure plots Polk County PILT
(dashed line) and Hiwassee Ranger District timber sale
receipts (solid line). Unit plan approved 1975 (arrow) and
implemented 1975-present. Impact classified as "spurious."

"Dispute": Hiwassee Unit Plan, Silviculture. The Hiwassee EIS described the silvicultural objectives of the plan as follows: "Even aged management will produce better quality timber in a shorter period of time with more economical harvesting practices and create less damage to the environment" (Forest Service 1975: 20). Debates over this objective raged throughout the comments reprinted in the Hiwassee FEIS, and were still smoldering eight years later.

Interviews revealed two levels of controversy, one within the forestry community and the other between the foresters and the public. First, some Forest Service officers and timber company managers disagreed with the forest supervisor's policy of programming large clearcuts in adjacent compartments.[15] The dispute involved the order of entry, that is, the sequence and location of timber sales. The Cherokee forest supervisor pursued a forest-wide policy of entering compartments in clusters because of cost efficiencies in building timber roads. The district ranger and some older timber-management staffers opposed this policy because the requirements for clustering led to logging of immature stands, where the trees had not reached the cumulation of annual growth that foresters consider optimum for logging. The new policy also meant that the Hiwassee EIS's order of entry (i.e., Forest Service 1975: Map C) was not followed during 1980-1983.

Second, interviews with recreation and environmentalist interests confirmed that the Hiwassee's timber management was also subject to the bickering about clearcutting southern mixed-hardwood stands that is common throughout the region. For example, a two-generation mountain family who were the principal merchants within the Hiwassee unit complained about "this major clearcutting spree," specific clearcuts affecting their property, overbuilding of timber roads (which led to some defecit sales, since road costs are credited against the stumpage price the Forest Service receives), the inequity of subsidizing high-standard timber roads when recreation roads were inadequately funded, and so forth. In the face of these controversies, the veracity of the EIS's vague, judgmental silviculture forecast is best classified as "impact disputed."

Unanticipated Consequences

One of the clearest purposes of NEPA's environmental assessment process is to force agency staffs to anticipate projects' consequences. This study's accuracy classifica-

214

tion scheme includes four separate codes for cases in which impacts were not properly anticipated. Two codes, "under-anticipated-adverse" and "underanticipated-beneficial," deal with cases in which EIS writers so vaguely or evasivly discussed an issue that our coders could not recognize the passage as a bona fide forecast. Four of the five "under-anticipated" cases were reviewed in the first section of this chapter. Two unanticipated impacts—cases in which the EIS contains not a hint of a possible consequence—are presented here.

Henrico County Wastewater Treatment Plant. This project involves a major new sewage treatment plant in eastern Henrico County (Richmond), Virginia. The proposal was opposed by a group of local residents who believed the system was designed to foster growth in the undeveloped eastern third of the county. One of the group's principal tactics was to prove that the proposed sewage treatment plant site was historically significant and thus should preserved. The so-called Deep Bottom site was the location of a Civil War battle after which 14 black Union soldiers were awarded the Congressional Medal of Honor, and the comments section of the Henrico FEIS (EPA 1978) is filled with dozens of pages of letters, battle maps, and other submissions detailing the significance of the site.

The local homeowners' campaign to have the plant site added to the Richmond National Battlefield Park failed. Park Service managers in Richmond interpreted the opponents' arguments as a tactical ploy and refused to be drawn into blocking this major federal-county project. Among other issues, adding a site to the park memorializing black Union troops would have irked the park's powerful constituency of Confederacy history buffs, who had established the park as a state facility before Virginia turned the land over to the National Park Service after the Depression. However, the controversy over the sewage plant brought the role of these troops to the attention of the park superintendent, one of very few black line officers in the federal land management agencies. Following completion of a historical survey and report, the park's managers added extensive information about the black Medal of Honor winners and Deep Bottom battle in the park's interpretive brochure and audiovisual displays, and incorporated the role of black troops in other park planning and interpretive documents. This serendipi-tous outcome, which was an impact of the EIS process rather than of the sewage plant per se, was classified as "unan-ticipated, but beneficial."

215

Tacoma Harbor: Resuspension of Toxics. As noted above, the most basic concern of environmental assessment is to avoid significant unanticipated adverse impacts. This concern, however, involves a fundamental dilemma: if an impact is truely unanticipated, then no one will probably perform the systematic monitoring needed to detect or document it. The problem of toxic substances in the sediments of Commencement Bay illustrates this dilemma.

The Tacoma harbor consists of eight parallel channels on the east side of Commencement Bay in southern Puget Sound. A March 1974 DEIS proposed "maintenance dredging" of one of the port's main channels, Blair Waterway, to a depth of 40 feet.[16] A month later, acting under a special CEQ dispensation, the Corps let a contract for the Blair Waterway dredging, and the dredging took place during June-September 1974.

A massive amount of sophisticated research has documented the fact that Commencement Bay contains a very foul soup of toxic substances, including synthetic chemicals, like chlorinated butadienes and polychlorinated biphenyls (PCBs), and trace metals like arsenic, lead, and mercury. These subtances enter the bay both through nonpoint-source runoff and atmospheric deposition from industrial plants around the bay and waterways emptying into the bay, notably the ASARCO copper smelter and a Hooker Chemical plant. The research also documents a variety of biological impacts the toxics have on the marine life in the bay.[17]

Blair Waterway is among the most polluted sites in Puget Sound. One of our interviewees, the marine biologist who managed the Puget Sound studies, forcefully argued that a dredging operation in Blair Waterway, such as that in 1974, would surely resuspend toxics that had settled in the waterway's sediments, thus exacerbating the toxic conditions in the waterway and Commencement Bay. Corps and Port of Tacoma interviewees did not dispute this point and, indeed, the Corps' generic assessment of its activities in Commencement Bay (Dames & Moore 1983: 5-2) acknowledges that dredging will cause some resuspension of sediment toxics. However, we cannot know the seriousness of any resuspension of toxicants caused by the 1974 Blair dredging. The Corps used a hydraulic dredge, which appears to have minimized the turbidity of the dredging operation (a separate impact audited during the fieldwork), and hence should have minimized any resuspension of toxics. In addition, the Corps' monitoring of its disposal site indicated relatively low concentrations of heavy metals in the dreging spoils effluent. However, no monitoring was done to detect resus-

216

pension of toxics during the 1974 dredging.

There is no evidence that anyone connected with the Tacoma dredging EIS was aware of this problem in 1974-1975. That is, there is no mention of the problem in the text of the EIS, and no comments on the DEIS even vaguely allude to the problem. Everyone in the Seattle-Tacoma water resources community now knows about the Commencement Bay toxics problem, but the first studies documenting the problem began in 1978. Indeed, the improvements in water chemistry surveying methods necessary to identify and document the Puget Sound toxics problem did not occur until the mid-1970s. Thus, we conclude that some resuspension of toxics was likely caused by the 1974 dredging, but that impact was not anticipated by the EIS writers. However, because the impact was unanticipated (and because the proper chemical monitoring tools were just becoming available in 1974), we do not have valid data indicating how adverse this impact was.

Inaccurate Forecasts

Most of the accuracy classifications cover cases in which the actual impact is not accurate enough to merit a "close" rating but there is a large enough a grey area that we have given the forecast the benefit of the doubt. The final classification, "inconsistent," was reserved for forecasts that appear to be flat-out wrong. This second half of Chapter 6 has provided only one to two examples of each classification. Overall, it thus contains representative numbers of major classifications, such as "close" and "range-vague". This subsection, however, presents five examples of the conceptually important least accurate forecasts.

Skipanon River Bridge: Upriver Development. This project is one of two in the sample that did not seem to involve either a major action or significant environmental impacts, as required to trigger the legal requirement for an EIS. The bridge is a 671-foot-long, two-lane replacement span located, on virtually the same alignment as its predecessor, in Warrenton, Oregon. The EIS was written because a local dredge operator, the instigator of this controversy, wanted a drawbridge built so that dredges and boats with masts could sail upriver. He then learned that the old bridge violated the terms of a 1916 Coast Guard permit, and induced the Coast Guard to issue a compliance order to the Oregon State Highway Division. The 1975 FEIS

Table 6.6
Skipanon Bridge, upriver development: Building permits, Warrenton, Oregon, for blocks adjacent to Skipanon River. Forecast: "this project would . . . encourage residential development along the river above the existing bridge, due to water-based amenities." Construction began December 1977; new bridge completed June 1979 (arrow). Impact classified as "inconsistent" with forecast.

Year	No. Building Permits		$ Value of Permit Work	
	New Bldg.	Remodeling	New Bldg.	Remodeling
1983	0	1	0	$2,000
1982	$\frac{1}{2}$[a]	1	$4,250[a]	$8,000
1981	0	0	0	0
1980	0	3	0	$15,500
->1979	$\frac{1}{2}$[a]	1	$4,250[a]	$6,000
1978	0	0	0	0
1977	0	0	0	0
1976	0	0	0	0
1975	0	0	0	0
1974	2	0	$27,000	0
1973	0	0	0	0
1972	0	0	0	0
1971	0	0	0	0
1970	0	0	0	0

[a] Same permit recorded in 1979 and 1982 for same work; permit apportioned equally between 1979 and 1982.

filed on the project proposed a new bridge but rejected the drawbridge demand. The proposal was the subject of a year-long political battle and town special election.[18] The Oregon Highway Department began construction in December 1977 and opened its new fixed-span bridge in June 1979.

The dredge operator wished to enhance access to and hopefully spur development of his property upriver. The Oregon State Highway Division (1975: 27, 34, 48, 59) mentioned this potential impact unusually often throughout its FEIS: "A secondary impact of this project would be to encourage residential development along the river upriver above the existing bridge, due to water-based amenities," "The anticipated increase in navigation will result in pressure for water-oriented residential development along the Skipanon River shoreline," " . . . long-term alteration of upstream river use patterns," " . . . may result in more intensive use of the river," and so forth.

As shown in Table 6.6, virtually no development occurred on the Skipanon River after the project. Actually, new upriver building declined by half during the postproject period, from two new houses to one. Postproject development was limited to four remodeling jobs worth a total of $25,500. Upriver development did not boom because there was no good reason for it. Warrenton is a small, one-stoplight town. Two large marinas operated downriver of the Skipanon bridge, and by 1983 one had fallen into bankruptcy. Moreover, the new bridge is too low to permit easy navigation by boats with high masts. The Skipanon upriver from the bridge is navigable for less than a half-mile because of an earthen tidal dam, and the east side of the river (the dredge operator's property) is largely wetlands. The forecast was classified as grossly "inconsistent" with the actual impact. Indeed, the EIS's frequent allusions to upriver development should be read as less a forecast than a reflex response to this local tempest in a teapot.

Weymouth-Fore and Town Rivers: Shipping Drafts. Economic benefits are often the key objectives of projects. Economic forecasts tend to be less accurate than other forecasts, and the main economic objective of the Weymouth-Fore project illustrates an outcome that is inconsistent with its forecast. The forecast states that the channel work would "allow deeper draft tankers" to use the Weymouth-Fore and Town rivers harbor. This is the key assumption in the cost-benefit analysis justifying the project, not a minor forecast. Corps economists assume that the same gross tonnage will be shipped through the harbor, with savings resulting from a shift in tonnage from inefficient, shallow-draft hulls to fewer, deeper-draft vessels. Thus, one would expect an increase in 30-foot-or-deeper-draft inbound vessels (outbound vessels become shallower after unloading) from 1977 onward, and a corresponding decrease in shallow tankers.

Actually, the reverse occurred. After the channel was deepened beyond its original authorized depth of 28 feet in 1977, the number of 30-foot-or-deeper ships decreased from the 1969-1975 range of 30-50 to 15-25 during 1978-1981 (Figure 6.21, solid line) and the number of shallow, 18-

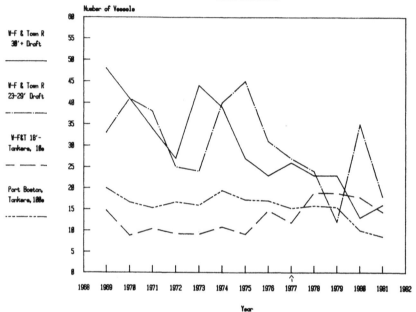

Weymouth-Fore: Shipping
Inbound Vessel Drafts

Number of Vessels

W-F & Town R
30'+ Draft

W-F & Town R
23-29' Draft

W-F&T 18'-
Tankers, 18+

Port Boston,
Tankers, 18+

Year

Figure 6.21
Weymouth-Fore and Town Rivers Harbor shipping efficiency:
vessel drafts of inbound cargo ships, including deep-draft
vessels (30-foot or greater draft, solid line), shallow-
draft tankers (18-foot or less, dashed line), and all Port
of Boston tankers. Forecast: "[The project] will allow
deeper draft tankers, without lightening their load or being
subject to tidal delays, to deliver their cargo to the Town
River Terminal." The shift from shallow, inefficient
tankers to deeper-draft tankers is the primary benefit of
the project in the Corps' B/C analysis. Channel work
completed April 1977 (arrow). Impact classified as
"inconsistent" with forecast.

220

foot-or-less tankers increased. This occurred because the principal customer for deep-draft tankers closed at the same time the channel was deepened. Boston Edison's Edgar Station coal/oil wharf is located on a large natural turning basin at the confluence of the Weymouth-Fore and Town rivers. Before the 1974-1976 channel work, 30-foot-or-deeper tankers could sail up to the Edgar Station wharf on high tide, offload as the tide receded, and then take advantage of the large turning basin to sail out. However, the Edgar Station closed in the wake of the oil-to-coal conversion policy of the mid-1970s. By 1982 tanker destinations were limited to Citgo, Mobil, and Quincy Oil gasoline tank farms upriver, where access is difficult except for shallower tankers and barges.

Ozone Unit Plan: Grazing. This planning unit covers approximately half a ranger distict on the Ozark National Forest in northwest Arkansas. The unit plan, which was administratively approved in the fall of 1976, generally contemplated standard multiple-use management of the unit's timber, mineral, range, and recreation resources. The Forest Service's (1976b: 89) list of target output levels in the Ozone unit plan FEIS projected an increase in livestock grazing use from 1,089 to 2,401 animal unit months.

Ozone unit grazing permits actually fell precipitously after the unit plan was implemented. After averaging 1213 AUMs during 1970-1975, grazing fell from 772 to 62 AUMs during 1976-1979, hitting zero by 1980. Basically, as explained by the district ranger, grazing allotments on the unit were not fenced, and Arkansas had passed and enforced a law prohibiting cattle from grazing loose. Demand for grazing permits withered because of the high cost of fencing, the steep topography of the unit, and the relatively small amount of national forest forage available. This outcome in the opposite direction from that forecast was also classified as "inconsistent."

Weyerhaeuser Road: Rock Painting. The EIS on this project is the second in the sample that does not seem to meet the criteria for writing an EIS. The road is located in the area of intermingled sections of federal and private lands in the Cascade Mountains east of Seattle, Washington. The proposal involved a 600-foot road right-of-way across Snoqualmie National Forest land to provide access for logging operations by the Weyerhaeuser Company on its private timberland. The right-of-way allowed an environmentally preferable road design, and Weyerhaeuser could have cut a road on its own land had the Forest Service rejected the proposal. Following Forest Service approval, the road was constructed in 1975. As the current ranger said during a fieldwork interview, "I was kind of awestruck that we would do an EIS on a rinky-dink project like this."

The large clearcuts Weyerhaeuser planned on its own land would have been made within the "visual impact zone" paralleling Interstate 90, so aesthetic issues dominated the EIS. The Forest Service (1974a: 8) thus proposed the following mitigation:

> Disturbance of the talus slide will expose unweathered rock surfaces without moss or lichen cover. A noticeable color contrast may result between disturbed and undisturbed rocks. The amount of color contrast will be reduced by the application of a coloring agent [a dilute asphalt emulsion] which adheres to the rocks and darkens them.

This mitigation struck us as so unusual that we included it on our list of fieldwork candidates. Weyerhaeuser's district timber supervisor, who had managed the project area since the FEIS, said the technical feasibility of such rock painting was unknown, as it had never been done before, such conditions were not included in the company's master cooperative agreement with the Forest Service, and the rock painting had definitely not been done. (None of the Forest Service officials interviewed for the study admitted to any knowledge about this mitigation.) This mitigation forecast was thus classified "inconsistent" with the actual outcome, with a beneficiality code of "trivial."

Newington Forest Subdivision: Firing Range. Interview and documentary information often provides imperfect evidence, particularly about social impacts. One case in the field sample, however, illustrates an impact for which such information is perfectly reliable. Newington Forest subdivision was built between 1977 and 1982 in suburban Fairfax County, Virginia, south of Washington, D.C. The subdivision is located just north of a police academy firing range on the grounds of the District of Columbia's Lorton prison. HUD's FEIS (1977: II-93) flatly asserted, in discussing public safety issues, that "there is no possibility that projectiles would overshoot the range."

This issue made the news immediately before our fieldwork trip. A military police contingent shot quite ineptly in April 1983 and peppered several Newington houses and yards with a few dozen 45-caliber bullets. The Newington Civic Association was up in arms, the U.S. District Court enjoined the D.C. Department of Corrections, and the local congressman attempted to impound the federal D.C appropriations, according to a numerous interviewees and five stories in the Washington Post. The forecast is classified as "inconsistent" with the actual impact.

SUMMARY

Each of the cases reviewed above involves unique details and nuances. These case studies demonstrate that a wide range of factors affect the accuracy of EIS forecasts and that most forecasts cannot be simply classified right or wrong. This chapter concentrates on the full variance in the case studies' forecasts, impacts, and accuracy classifications. Indeed, if we have succeeded in honestly and fully explaining these cases, with their grey areas, critical readers may have occasionally disagreed with a classification. Thus, in a sense, a thorough Chapter 6 would have to explain all 239 audited cases.

Chapter 7, in contrast, will summarize the distributions of accuracy classifications and examine their relationship to other characteristics of the EISs in the field sample. The cases reviewed above nonetheless presage the main features of the overall distribution. The classifications that occurred often among the Paint Creek, Sequoyah, Grand Teton, and Cleveland harbor cases are also the most common classifications overall--the very accurate, "close" rating and the very equivocal "impact within the range of vague forecast" rating.

NOTES

1. This chapter reviews 58 of the 239 cases in the study. For comparision with later analyses (e.g., Table 7.1), the numbers of example accuracy classifications, together with the accuracy match abbreviations used in Tables 6.1, 6.3, 6.4, and 6.5, are as follows:

Primary Match Code	Full Specification	No. Cases Reviewed in Ch. 6
Close	Close (i.e., accurate)	14
Complex+	Complex, but basicly accurate	3
Dispute	Impact disputed by informants	1
Exceeds	Impacts exceeds that forecast	4
Less	Impact less than that forecast	4
Incons.	Impact inconsistent with forecast	7
NIm/NFor	No clear impact, no-impact forecast	3
NIm/Some	No clear impact, some impact forecast	2
Not yet	Impact has not yet occurred	1
Obvious	Accuracy intuitively obvious	1
Range-VF	Impact within the range of vague forecast	9
Spurious	Impact wholly spurious	2
Unant+	Unanticipated impact, but beneficial	2
Underant+	Underanticipated impact, but beneficial	1
Unant-	Unanticipated impact, adverse	1
Underant-	Underanticipated impact, adverse	3

2. Data sources are not cited in this chapter, except for a few systematic scientific studies. The sources for Figure 6.1 and Table 6.2, for example, are the U.S. Geological Survey's annual Water Resources Data, Ohio. See Chapter 5 for a description of the study's data sources.

3. The Corps operating target release rate for Paint Creek dam is a maximum of 5,900 cfs during the December-April peak flow period (Corps of Engineers 1974b: 5). Eight of 40 December-April monthly streamflow maxima exceeded this target during the 1974-1975 to 1981-1982 water years.

4. The skimpy data for this case (generally only one or two observations per year) are not graphed in a separate figure. Several other cases' time-series data are also not graphed in the interest of conserving space.

5. Essentially, the curve graphing temperatures (horizontal) at reservoir depths (vertical) turn or change slope twice. The zone between the turns is the thermocline, that is, the zone between the warmer upper stratum and the colder lower stratum. The BASIC program identifies the depths of the turns. The anaerobic depth can be estimated directly from the Corps' printout. The drop from DO > 1.0 mg/l to DO = 0 mg/l is usually discontinuous.

6. Chicago Tribune (January 5, 1986: 5; January 6, 1986: 3), and other newsclips.

7. Plant officials provided the field crew with the exact percentage of nonstandard feedstock, but asked that

this proprietary information remain confidential.

8. Carlisle House, an old stagecoach station, was pre-served and moved to a better site a mile north of the plant. While the EIS's statement did not qualify as a forecast, the field crew inspected the new Carlisle House site and veri-fied that it is indeed managed as a public attraction.

9. At the time of the fieldwork, a major controversy about national park policy involved Secretary of the Interior Watt's scheme to use funds reserved for purchasing park land, such as inholdings, for structural projects like sewage plants (Culhane 1984: 308-310). Our interviewees' quibbles thus may have reflected a general in-service disagreement with Secretary Watt's policy.

10. In particular, the EIS lacks a clear statement of the facility's objective, aside from the obvious function of containing dredging spoils. Interviewees suggested that an unstated objective of site no. 12 was to provide landfill for expansion of Cleveland's lakefront airport. The EIS contains a sentence, buried deep in the document, which our coders treated as a nonforecast: "It is contemplated that the area will be offered to the airport authority for future use in connection with airport expansion." After site no. 12 was built, city bureaucrats competed over the loca-tion of the next disposal site. Site no. 10, together with site no. 12, would have allowed the airport to expand its runways. Site no. 14, two miles east, would provide erosion protection for a city park, and site no. 14 prevailed. That is, site no. 12 was indeed made available to the airport, but the critical parcel needed by the airport was not filled, so the expansion did not occur. Because we could not draw a sensible conclusion about the match between this outcome and the EIS's nonforecast, this case was excluded from the audit results reported in Table 6.5.

11. Apart from the routine nature of the project itself, the poor audit information on disposal site no. 12 seems to be simply bad luck. The Cleveland fieldwork took place on our first full-fledged field trip. An earlier training project in the Chicago area (Illinois Beach) yielded average impact data and the other projects on the first field trip, Paint Creek and the Shippingport breeder reactor, yielded excellent audit information. While the Corps' Huntington district provided excellent data on the Paint Creek dam, we were unimpressed by the Corps' moni-toring of the three dredging projects in the sample.

12. Personal communication, Orie Loucks, May 26, 1987.

13. The "complex but apparently inaccurate" classifi-cation was assigned to two cases in which interview infor-

mation could not be reliably confirmed. For example, in a Skipannon bridge case, interviewees claimed a wildlife coordination had taken place but their agencies' files suggested the coordination may not have occurred.

14. Source: Statistical Handbook on Aviation, FAA, 1975-1984, cited by Robert Grierson in a paper on airline deregulation, Public Administration 420, Northern Illinois University, December 1986.

15. Compartments are the smallest areas of land in Forest Service planning and management. Clearcutting or even-aged management consists of cutting all trees in a given area and then replanting the stand with the same species of tree, which will then mature at the same time.

16. The Blair Waterway component of the project seems suspiciously related to ongoing Tacoma harbor developments. The action was described as "maintenance dredging," but the channel's authorized depth in 1974 was only 35 feet. The EIS attributed three feet of the extra dredging to 40 feet as "advanced maintenance" and two feet as "allowable over-depth." However, a January 1977 Corps FEIS proposed deepening Blair Waterway to 45 feet, plus deepening nearby Sitcum Waterway to 40 feet. This 1977 proposal was introduced in Congress in 1979, but no authorization had passed as of 1983. Thus, the 1974 dredging suggests not "advanced maintenance," but incremental channel deepening.

17. This research includes a systematic series of studies of toxic concentrations (e.g., Dexter et al. 1981, Riley et al. 1981) and biological effects (e.g., Malins et al. 1982, Chapman et al. 1982) by National Oceanographic and Atmospheric Administration oceanographers in Seattle, plus a survey of toxic sedements (e.g., Johnson et al. 1983) by the Washington State Department of Ecology. The Corps' "COBS II" report (Dames & Moore 1983) accepted the basic veracity of these findings. The problem of toxicant resuspension in Commencement Bay has also received some national notoriety within policy circles (Stanfield 1985: 1879).

18. We do not wish to denigrate either the Skipanon EIS or the new bridge. This EIS contains a thorough record of project decisionmaking. The old bridge was also a very outmoded "float-out" design. If a vessel needed to sail upriver, pontoons were needed to float the bridge out of the way on a high tide. This procedure (which the operator reportedly requested a dozen times during the controversy), disrupted both Warrenton's highway access and the municipal water supply, since the town water main was suspended under the bridge. Thus, the Coast Guard order provided an opportunity to execute a necessary improvement.

226

7
Forecast Accuracy: A Summary

This chapter provides a summary answer to the fundamental question of our evaluation of U.S. environmental assessments--are EIS forecasts accurate? The complicated rating scheme derived at the end of Chapter 5 and the case studies in Chapter 6 have prepared for the conclusion that the EIS forecasts in the study's sample are not completely accurate, or at least not uniformly so. Nonetheless, the overall distribution of ratings in the set of 239 forecast impacts contains both good and bad news about the accuracy of U.S. environmental assessments.

A second objective of this chapter is to analyze factors that may help explain the variance in environmental forecast accuracy. One issue we examine is whether, as cynical critics of EISs commonly suspect, EIS writers tend to overstate project benefits and understate adverse impacts because of the adversarial, political nature of the EIS process. We also examine whether, as rational systems analysts might expect, any relationship exists between the technical precision of forecasts and their accuracy. Third, within the limitations of the information gathered under our research design and fieldwork, we explore whether several institutional-political characteristics of EISs are related to predictive accuracy. In general, however, our analysis reveals few interesting explanations for the overall distribution of predictive accuracy ratings in the sample. Thus, that distribution itself probably constitutes the major finding of our audit.

Table 7.1
Direction of field-sample impacts relative to forecast direction. "Secondary direction" refers to secondary codings of the forecast accuracy of a given case; the "secondary" column includes 47 cases with a secondary direction code and five cases with both secondary and tertiary direction codes. N = 239 forecast impacts.

Direction of Impact Relative to Forecast	Primary Direction	Secondary Direction
Correct	170	31
Wrong	21	3
Impact Direction Indeterminate		
No significant postproject change	15	3
Continuous preproject/postproject trend	14	1
Variance in indicator much greater than any postproject change	12	9
Nondirectional impact, "bad"	4	3
" " , "good"	2	2
Uncodable (impact disputed)	1	
Totals	239	52

THE DISTRIBUTION OF ACCURACY CLASSIFICATIONS

The most rudimentary requirement of an accurate forecast is that the actual impact be in the same direction as predicted. Almost three fourths of the impacts audited in the field sample are in the same direction as forecast, and only 21 impacts are in an unambiguously wrong direction. No clear change is visible in most of the rest of the cases: a time series may show a fairly constant increase or decrease, for example, or the natural year-to-year or other periodic variance in the impact's indicator may be much greater than any change in the average value of the indicator before and after the project. A half-dozen impacts were coded as nondirectional; for example, the I-94 interchange cost more than forecast, but the direction of that cost overrun cannot be coded as either correct or incorrect, only as "bad." Fifty-two cases received secondary impact direction codes. (Readers are reminded that the audit's unit of

228

analysis, a _forecast impact_, covers a single type of impact. A case could receive secondary direction, accuracy match, or benefiiality codes to cover minor differences in classification attributable to secondary data or interview information or to secondary forecasts dealing with the same impact.) As shown in Table 7.1, the distribution of secondary direction codes is essentially the same as for the primary distribution. Thus, the first "good news" about EISs is that most forecasts at least prove to be in the correct direction.

A second bit of good news is that the most accurate classification of the match between forecasts and actual impacts--"close"--is the modal rating in the study. (See Table 7.2, column 1.) The Paint Creek limnology, Sequoyah property taxes, Grand Teton fire management, and Weymouth-Fore blasting cases, reviewed in Chapter 6, provide good examples of this rating. Four more cases, among them the Grand Teton wilderness example, are "complex" but basically accurate in our judgment. Thus, 30% of the sample forecasts were rated as fairly accurate.

Third, very few forecasts in the sample are clearly inaccurate. Some forecasts are wrong, and the reasons for the mistaken forecasts are sometimes fascinating. These cases include the Paint Creek water supply, Sequoyah fluorides, Skipanon upriver development, Weymouth-Fore shipping, and Newington firing range cases described in Chapter 6. But these cases are among only 15 impacts classified as "inconsistent" with their forecasts.

Fourth, field interviews with knowledgeable informants revealed only three actual impacts that had not been explicitly anticipated by EIS writers, and two of these were minor serendipitous outcomes. The only adverse unanticipated impact in the study, the Tacoma harbor toxics resuspension case, can be inferred only because studies subsequent to the EIS revealed the toxicity of the harbor's sediments. Another five impacts were so understated by EIS writers that they could not be considered properly anticipated. Four underanticipated impacts were adverse and three were so understated that our coders did not recognize them as bona fide forecasts during the content analysis. One underanticipated adverse impact, the Cleveland harbor power plant operation case, involved a flat denial by the Corps of an EPA comment that proved to be correct, and the other three leave the impression that EIS writers fudged inconvenient impacts.

Identifying potential types of impacts is perhaps more critical for good environmental assessment than precisely predicting the magnitude of impacts. Subtle unanticipated

Table 7.2
Summary classification of matches between EIS forecasts and actual postproject impacts. The secondary-match column includes 89 cases with secondary match codes and nine cases with both secondary and tertiary codes. Accuracy classifications are grouped into four similar ranks containing comparable accuracy classifications. Percents total more than 100% because of rounding.

Impact-Forecast Match Classification	Primary Match	Secondary Match	Totals No.	Totals %
Match Rank = 4				
CLOSE	67	1	68	20.2
COMPLEX, but arguably accurate	4	9	13	3.9
MITIGATION measure DONE	0	15	15	4.5
Match Rank = 3				
Impact within the RANGE of VAGUE forecast	64	7	71	21.1
NO clear impact, NONE forecast	11	4	15	4.5
Accuracy INTUITIVELY OBVIOUS	5	2	7	2.1
Impact has NOT YET occurred	4	6	10	3.0
Match Rank = 2				
Impact LESS than forecast	24	3	27	8.0
Impact EXCEEDS forecast	10	9	19	5.6
NO clear impact, SOME impact forecast	18	3	21	6.2
COMPLEX, essentially inaccurate	2	8	10	3.0
Apparent impact SPURIOUS	5	12	17	5.0
Impact DISPUTED	2	3	5	1.5
UNANTICIPATED, but beneficial	2	1	3	0.9
UNDERANTICIPATED, but beneficial	1	0	1	0.3
Match Rank = 1				
INCONSISTENT	15	12	27	8.0
UNDERANTICIPATED, adverse	4	1	5	1.5
UNANTICIPATED, adverse	1	0	1	0.3
MITIGATION measure NOT DONE	0	2	2	0.6
Totals	239	98	337	100%

impacts pose a major problem for environmental auditors. Because reasonable people, like EIS writers, do not antici- pate problems beforehand, a problem is likely to remain unseen after a project is completed simply because no one is looking for it. For example, since the Tacoma Harbor toxicant problem was unanticipated, no monitoring was done that could document the magnitude or significance of any resuspension caused by dredging.[1] It is, therefore, quite possible that more than three unanticipated impacts occurred among our 29 projects that neither we nor anyone else know about. Nonetheless, finding almost no evident unanticipated impacts and very few underanticipated impacts is probably the best news about EISs to emerge from our audit.

The bad news about EISs is that two thirds of the forecasts we audited fall into a grey area between accuracy and clear inaccuracy. The second most common classification describes impacts "within the range of vague forecasts." This group of impacts, whose accuracy is confounded chiefly by their forecasts' imprecision, falls only three cases shy of the sample mode and accounts for over a quarter of the primary accuracy ratings. Two other classifications are also accurate because of intrinsic (if understandable) properties that make the forecast hard to prove wrong. Five forecast impacts are logically obvious results of the exe- cution of the proposal (e.g., Chapter 6's Illinois Beach habitat preservation case). Any random sample of EIS forecast impacts would contain more than the 2% of these ratings shown in Table 7.2, but our field data-gathering criteria discouraged the selection of incidental-impact forecasts because they are generally uninteresting. Another four cases received "not yet" ratings, which indicate these forecasts allowed for the possibility of long-term impacts that might not have occurred by the time of our field audit.

A half-dozen ratings cover forecasts that are not very accurate but cannot be classified as clearly inconsistent with the actual impact. The "less" and "exceeds" classifi- cations, which are logically equivalent ratings, account for 15% of the sample's forecast impacts. That is, these fore- casts are in the correct direction, but the magnitude of the impact was significantly less or greater than forecast. The "no clear impact, some impact forecast" classification accounts for another 8% of the cases. (Recall that these ratings connote "no clear impact" because, for example, the data showed wide preproject, postproject variance or con- tinuous preproject, postproject trends.)

As shown in Table 7.2, 89 cases received more than one accuracy-match code. Two accuracy classifications are only used as secondary ratings--"mitigation done" and "mitigation not done." These codes describe cases, like the Weymouth-Fore blasting safety case, in which the primary classification covers a project impact and the secondary classification covers a mitigation measure (i.e., monitoring) dealing with the impact. In addition, a relatively large number of cases recieved "spurious" secondary ratings [2]; that is, the accuracy of 5% of the field sample's forecasts is wholly or significantly determined by causes unrelated to the EIS's project. Overall, however, the distribution of total accuracy ratings (i.e., primary-plus-secondary match codes) is quite similar to that of the primary ratings--a bimodal distribution with large numbers of "close" and "impact within the range of vague forecast" ratings, and only a small number of clearly incorrect forecasts.

A Rank-Order Accuracy Index

The distinct nominal accuracy classifications fall into a rough hierarchy, as indicated by the four ranks in Table 7.2. Some classifications fall at the same rank because of clear conceptual similarities, such as the "unanticipated-adverse" and "inconsistent" classifications. The "complex-inaccurate" and "impact disputed" ratings are essentially the counterparts for ratings based on verbal information to the "exceeds" and "less" ratings that can be given based on quantified impact data. Other ratings are grouped together because of similarities observed during our audit. A dozen "no clear impact, none forecast" cases' forecasts tend to be vague, intuitively obvious, or trivial, so they are ranked with the "impact within the range of vague forecast" and "intuitively obvious" codes.

Other classifications are listed a rank higher than they might have been since we believed evaluators must prove a forecast to be inaccurate. For example, cases rated "not yet" are inaccurate in that the predicted impact did not occur, but we give the case the benefit of the doubt since the forecast allows for the possibility that the impact may still occur in the future. The serendititous "unanticipated-beneficial" and "underanticipated-beneficial" seemed like lesser errors than their adverse and generally serious counterparts. Finally, in a case classified as "spurious," the actual impact is generally inconsistent with the forecast but the case deserved a more generous rating because its inaccuracy was unrelated to the EIS or project.

The four classification ranks form a natural ordinal ratings index. A case usually receives the ordinal value associated with its primary match classification. However, certain cases received an increment or decrement of one when their secondary classifications were significantly higher or lower than the primary match. (See the methodology appendix for further information about this index.) The index provides a rough numerical measure of forecast accuracy, which is useful because the 17 primary accuracy classifications do not lend themselves to ready analysis. The mean index score across the sample of 239 cases is 2.8--that is, slightly above the midpoint of the range of accuracy ranks.

Forecast Accuracy and Case Characteristics

Mitigations. The most systematic pattern in predictive accuracy in the field sample is that mitigation promises are generally kept. As shown in Table 7.3, the 35 mitigations in the sample have an average accuracy rating of 3.28, with 23 classified as "close." Only five mitigations were not carried out, and the adverseness in three of those cases is classified "trivial." The Weyerhaeuser Company's failure to paint the rocks exposed by its road cut, as suggested by the Forest Service (1974a), provides an apt example of these trivial, "inconsistent" mitigations. Moreover, 15 impacts received secondary ratings of "mitigation done," versus only two "mitigation not done."

Basically, mitigations involve fairly discrete actions that can be directly managed by the lead agency or project sponsor. Once incorporated into the project plan, these mitigation measures are implemented as regularly as are other aspects of the plan. The only exception to this generalization seems to involve monitoring, which might need to be conducted after the project is completed. Three of the five monitoring mitigations were not carried out; only two of 20 ameliorative measures, which are generally done during the project-execution phase, were unfulfilled. In any case, EIS statements about mitigations are more straightforward than forecasts about future impacts affected by various causes, many of which are out of the control of project managers. So higher accuracy among the mitigation cases should be expected.

Substantive Category. There is little variance in the overall average accuracy ratings of the four substantive categories of forecasts. (See Table 7.3, column 4.) Social

Table 7.3

Average predictive accuracy broken down by primary characteristics of field sample cases. Cell numbers are means of the accuracy ordinal ranks as shown in Table 7.2--that is, "4" for "close" cases, "3" for "impact within the range of a vague forecast," "1" for "inconsistent" cases, and so forth. Numbers of forecast impacts are in parentheses. The standard deviation of the accuracy ordinals for the whole sample of 239 cases equals 0.942.

Category	Impacts	Objectives	Mitigations	Totals
Physiographic	2.67 (N=28)	3.20 (N=10)	3.07 (N=14)	2.88 (N=52)
Biological	2.61 (N=21)	2.25 (N=4)	3.62 (N=8)	2.81 (N=33)
Economic	2.61 (N=49)	2.18 (N=16)	(N=0)	2.50 (N=65)
Social	2.91 (N=61)	3.06 (N=15)	3.30 (N=13)	3.00 (N=89)
Totals	2.74 (N=159)	2.71 (N=45)	3.28 (N=35)	2.816 (N=239)

forecasts are somewhat more accurate than the other forecasts. Economic forecasts have a lower average accuracy, but the 0.3 difference between the economic forecasts' and the sample's means is fairly small compared to the sample's standard deviation of 0.94. Physiographic and biological cases have essentially the same average accuracy, at the sample mean.

These means mask a few interesting differences in accuracy classifications within the four categories. Economic objectives averaged a notably low 2.18 accuracy, with only one "close" rating among the 16 cases. In addition, as shown in Table 7.4, economic forecasts frequently fell into the "grey area." Almost twice as many economic impacts were rated as "within the range of a vague forecast"

234

Table 7.4
Impact-forecast matches within the four substantive impact categories. Entries are <u>primary</u> match classifications only. Accuracy classifications are grouped into four similar ranks, as in Table 7.2.

Impact-Forecast Match, Primary Classification	Physiographic	Biological	Economic	Social	Totals
Match Rank = 4					
CLOSE	20	9	9	29	67
COMPLEX, but arguably accurate	1	0	1	2	4
Match Rank = 3					
Impact within the RANGE of VAGUE forecast	12	8	16	28	64
NO clear impact, NONE forecast	1	2	3	5	11
Accuracy INTUITIVELY OBVIOUS	0	2	0	3	5
Impact has NOT YET occurred	1	0	3	0	4
Match Rank = 2					
Impact LESS than forecast	5	3	8	8	24
Impact EXCEEDS forecast	1	0	6	3	10
NO clear impact, SOME impact forecast	1	3	11	3	18
COMPLEX, essentially inaccurate	1	1	0	0	2
Impact DISPUTED	0	2	0	0	2
Apparent impact SPURIOUS	0	1	3	1	5
UNANTICIPATED, but beneficial	1	0	0	1	2
UNDERANTICIPATED, beneficial	0	0	1	0	1
Match Rank = 1					
INCONSISTENT	5	1	3	6	15
UNANTICIPATED, adverse	1	0	0	0	1
UNDERANTICIPATED, adverse	2	1	1	0	4
Totals	52	33	65	89	239

as were rated "close"; 38% of the economic impacts were placed in the relatively inaccurate "less," "exceeds," and "no clear impact, some impact forecast" classifications, but only 15% of the other categories' cases received these ratings. This finding seems consistent with the conventional wisdom (e.g., Duinker 1985) that economic justifications for projects are usually overstated or inaccurate.

As Table 7.4 also shows, social impacts achieved their relatively high average with concentrations in the modal classifications, "close" (33% versus 25% for the other three categories) and "impact within the range of a vague forecast" (31% versus 24% for the other categories). By contrast, disproportionate numbers of physiographic cases are classified at the opposite ends of the rating scheme: 38% of the physiographic cases received "close" ratings (versus 25% for the other three categories), notably more physiographic cases received "close" than "range . . . vague" ratings, and 17% (versus 6% for the other categories) received "inconsistent" or "un/underanticipated" ratings.

Impact Types. More variance can be found among specific types of impacts than among the four main categories of cases. The average accuracy of the 21 impact types with more than five cases in the field sample ranges from 3.37 down to 1.60. (The impacts listed in Table 7.5 generally combine more than one content analysis impact-type code. For example, the noise and blasting vibration codes are combined because of similarities in their physiographic mechanisms and social effects.) There are no noteworthy patterns among the impact types with too few cases to merit listing in Table 7.5. In particular, no systematic accuracy patterns are evident among the unique forecasts in the sample. For example, Table 7.5 lists the unusual impact types with only one case per impact type, including the Jackson enplanements, Sequoyah process diversity, Newington firing range, and Grand Teton wilder-ness cases discussed in Chapter 6. These 17 cases have an average accuracy of 2.65, essentially the same as the average for ordinary impacts.

Several relatively accurate impact types involve the primary subject areas of important EIS-writing agencies in the sample. The dozen radiation forecasts are tied with traffic volume forecasts as the second most accurate in the sample. Forestry forecasts are also fairly accurate, although their 3.0 average masks a notable difference between three economic forecasts about timber production (accuracy = 3.67) and six biological forestry forecasts (accuracy = 2.67). In other words, the NRC, FHwA (actually

Table 7.5
Accuracy of forecasts among comparable impact types.
Similar impact-type codes (see Table 5.2) are combined
below; the numbers of distinct impact-type codes are shown
in brackets. Means are averages of the accuracy ordinal
ranks as shown in Table 7.2—that is, "4" for "close" and so
forth.

Impact Types [No. Codes]	X̄ Accuracy	N
Noise and blast vibration [2]	3.37	8
Radiation and radwaste [3]	3.33	12
Traffic volume (ADTs) [1]	3.33	9
Hydrology [3]	3.17	6
Erosion [1]	3.16	6
Aesthetics [2]	3.14	7
Aquatics (inclu. benthics) [7]	3.00	13
Forestry [3, Bio. + Econ.]	3.00	9
Population and demographics [2]	3.00	7
Recreation [6]	2.82	17
Employment [2]	2.78	9
Displacement, property owner [3]	2.75	8
Taxes [2]	2.71	14
Complex, unique impacts [17][a]	2.65	17
Development, general [3]	2.67	6
Traffic, other (non-ADT) impacts [4]	2.44	9
Water quality [6]	2.38	13
Agriculture [3]	2.29	7
Air quality [3]	2.29	7
Wildlife [3]	2.17	6
Energy production [2]	1.60	5

[a] These 18 impact types are airline enplanements, air
traffic safety, barge traffic, building industry stability,
engineering process diversity, firing range safety, fly ash
disposal, forest blowdown, housing quality maintenance,
police demand, public transit use, social services, R&D
engineering, transportation planning, scientific research,
soil contaimination, and wilderness designation.

the state highway departments), and the Forest Service forecast fairly accurately about impacts directly related to their primary missions.

The noise/vibration forecasts have the most accurate average in the sample. The two blast-vibration cases in this set take the prize for the best single impact type in the study, with ideal quantified forecasts, professional monitoring data, and clearly accurate forecast impacts. The noise cases per se achieve their high accuracy chiefly because of three "close" rated mitigations.

However, some impact types with high average accuracy ratings benefit from a quirk in our rank-order scheme. A case can receive an above-average ranking either by being very accurate (e.g., "4" for "close") or by having a fuzzy impact or forecast (e.g., a "3" for an "within the range of a vague forecast"). The seven forecasts about project aesthetics received two "close" ratings plus three "range . . vague" ratings that are natural consequences of the inherent subjectivity of forecasts about visual aesthetics. For example, AEC's EIS writers stated that the Monticello nuclear station would have an adverse visual impact because the plant's architecture was ugly. Aesthetic impacts can be quantified, but this forecast is phrased as a completely subjective judgement, although one with which the field crew concurred. Population forecasts' accuracy are almost entirely attributable to this quirk, with five of seven cases receiving "impact within the range of vague forecast" ratings. Murdock et al. (1982) discovered similar weaknesses in EISs' population forecasts.

Several impact types at the bottom of the accuracy distribution deserve more attention than some of the inflated impact types at the top of the distribution. Readers cannot be surprised that energy production forecasts are the least accurate type of impact--at more than one standard deviation below the sample mean--because the sample's projects were generally implemented during the instability in the industry after the energy crises of the 1970s. Three of these forecasts involve significantly overestimated electricity demand for midwestern power plants, and the other forecasts (an Ozone natural gas production forecast and the Cleveland harbor power plant operation case) were even worse. We are also not surprised by the poor accuracy ratings of the few wildlife cases in the sample. Despite the ubiquitous inclusion of lengthy species lists in EISs, agency managers generally have very poor information on wildlife. Thus, we could scrounge up only a half-dozen wildlife cases in the field sample, half of those cases' data were rated as "low quality," and the best accuracy classification in the bunch is a single "impact within the range of vague forecast" rating.

Of much greater concern, the air- and water-quality impact types occupy the third and fifth worst accuracy slots in Table 7.5. These are the two most commonly recognized types of environmental impacts, and the prescriptive literature on environmental assessment devotes disproportionate attention to predicting such impacts (e.g., Canter 1977: 49-119). As noted in Chapter 5, good air- and water-quality impact data are rare, so the reliabiliy of our subsample of these impacts is open to question. However, when we could get acceptable data, air- and water-quality forecasts were often disappointingly inaccurate. Only four of the 20 air- or water-quality cases received a "close" primary rating, two cases rated "close" were downgraded from a "4" to a "3" index score because of secondary ratings indicating relative inaccuracy, and the "close" air/water-quality cases were offset by four cases that received "inconsistent" ratings. In other words, the sample EISs' forecasts about these most fundamental types of impacts were very disappointing.

BENEFICIALITY AND ACCURACY

Most observers of NEPA have noted, more or less approvingly, the adversarial nature of the EIS process. (See, for example, Culhane 1974, Liroff 1976, Andrews 1976, Friesema and Culhane 1976, Orloff 1978, Fairfax and Andrews 1979, Mazmanian and Nienaber 1979, and Taylor 1984.) Draft EISs are subject to interagency, intergovernmental, and public criticisms, and this fairly channelized form of bickering overtops its banks into post-FEIS litigation often enough to remind people that the disputants are serious. Since the EIS process pits the proponent agency against potential adversaries, many critics of EISs assume that EIS writers overstate projections about the goals of projects and minimize adverse or controversial impacts.[3] Several indicators in our analysis have a bearing on this hypothesis and none support it.

First, if the benefit-overestimation hypothesis were correct, forecasts about project objectives might be notably less accurate than other kinds of forecasts. However, as shown in Table 7.3, forecasts about objectives averaged about as accurate as forecasts about ordinary impacts. As noted above, forecasts about economic objectives proved to be fairly inaccurate, but social and physiographic objec-

Table 7.6

Predictive accuracy compared with relative adverseness of impact forecasts. "Ambiguous beneficiality" forecasts are ones in which the beneficiality or adverseness of a predicted outcome depends on one's values; "neutral"-benefit forecasts are explicitly so described by the EIS. Means are averages of the accuracy ordinal ranks shown in Table 7.2.

Adverse/Beneficial Forecast	All Cases Accuracy	N	Excluding Mitigations Accuracy	N
Beneficial impact forecast	2.92	122	2.80	91
Neutral impact forecast	3.00	12	3.00	12
Ambiguous impact beneficiality	2.81	33	2.73	30
Adverse impact forecast	2.64	70	2.65	69
Uncodable (unanticipated impacts)	1.00	2	1.00	2
Totals	2.82	239	2.73	204

tives' forecasts were more accurate than average.

Second, the 21 cases in the field sample coded "most controversial" have the same accuracy average, 2.71, as ordinary, less controversial impacts and objectives.[4]

Third, forecasts of beneficial impacts are about as accurate as forecasts of adverse impacts. Beneficial-impact forecasts are slightly more accurate than adverse-impact forecasts across all 239 cases (Table 7.6, column 1). However, that small difference is an artifact of mitigation forecasts, which are invariably coded as "beneficial" (that is, as the avoidance of an adverse impact). When mitigations are excluded, the gap between the two types of forecasts' accuracy narrows to an insignificant 0.15.

Fourth, there is little difference in the beneficiality of actual impacts relative to their forecasts. As a third classification of forecast accuracy, in addition to direction and impact-forecast match codes, we recorded whether the actual impact was more or less beneficial or adverse than forecast. As shown in Table 7.7, a quarter of the cases were as beneficial or adverse as forecast, and 39% of the codes indicate that the impact could not be clearly

240

Table 7.7

Impact beneficiality. The codes describe the beneficiality of the actual impact relative to the forecast; for example, "more beneficial" means the actual impact was more beneficial than forecast, "as adverse" means the impact was as adverse as forecast. "Crossover-beneficial" means the actual impact was beneficial but had been forecast to be adverse. The four codes in the center column cannot be validly listed as either adverse or beneficial. Cell entries are numbers of forecast impacts, N = 239.

Beneficiality	N	N	N	Adverseness
More beneficial	15		28	Less adverse
Crossover-beneficial	2			
As beneficial	46		14	As adverse
As neutral		19		
Ambiguous		34		
Subjective		16		
Trivial		25		
Less beneficial	25		13	More adverse
			2	Crossover-adverse

classified as either adverse or beneficial. The key to interpreting Table 7.7 is that the four extreme corners of the table are almost perfectly symetrical. The 15 impacts more adverse than forecast (including two adverse impacts where benefits had been forecast) are offset by 17 impacts more beneficial than forecast, and the 25 "less beneficial" impacts are offset by 28 impacts less adverse than forecast. That is, the sample's EIS writers, if anything, underestimated benefits in about five cases.

ACCURACY AND FORECAST PRECISION

If the most cynical criticism of EIS writing finds no support in our findings, neither do the most rationalist hopes about precise environmental forecasting. Advocates of systematic, quantified EIS analysis clearly prefer that EIS forecasts be both technically proficient and accurate. Thus, they might logically expect that a forecast's technical precision would enhance its accuracy.[5] However, we found no real relationship between the technical precision of forecasts and their accuracy.

Forecasts' quantification is not related to predictive accuracy. As shown in Table 7.8, the few time-series forecasts average a high 3.33 accuracy ranking. This high average seems to result simply from three "close"-rated traffic volume cases (which qualify as "time-series" forecasts because FHwA invariably forecasts highway average daily traffic in an immediate postproject year and about a decade later); the other six time-series cases' ratings seem to relect merely good luck. Actually, the 13 forecasts that predict an impact falling within specified bounds on some indicator probably represent a better subset of cases. Statisticians would recognize such implicit "confidence intervals" to be the most sensible form of prediction, given the uncertainties of EIS forecasting, and these forecasts are relatively sound. In any case, the quantified forecasts as a whole are no more accurate than the rest of the cases. The 91 quantified forecasts average 2.81--the sample mean and essentially equal to the 2.84 average of the verbal, unquantified cases. The accuracy average of the verbal forecasts is, of course, inflated since almost all mitigation forecasts receive that code; excluding mitigations, the quantified forecasts average 2.79 and the verbal forecasts 2.70.

A forecast's use of explicit units of measurement is also unrelated to predictive accuracy. Forecasts that used economic and physiographic measurment units averaged only 0.07 and 0.02 better, respectively, than the means for all economic and physiographic forecasts.[6] The 38 forecasts that used social measurement units had an average accuracy of 2.91, exactly the same as the average for all social impacts, and the few cases that used biological measurement units, such as they are (i.e., acres and wildlife population counts), were actually less accurate, by 0.12, than average biological forecasts. Forecasts that employed no measurement unit at all have the same average accuracy as all ordinary impacts and objectives in the sample.

242

Table 7.8
Predictive accuracy compared with indicators of forecasts'
precision: forecasts' quantification, measurement units, and
certainty. Means are averages of the accuracy ordinal ranks
as shown in Table 7.2.

Forecasts' Characteristics	X̄ Accuracy	N
Quantification		
Quantified		
Time series	3.33	9
Single-number forecasts, multiindicator	2.80	20
Single-number forecast, 1 indicator	2.69	46
Bounded-values forecast	3.00	13
Percents (for noninterval measures)	2.33	3
"No impact" forecast	2.73	15
Verbal, unquantified forecast	2.84	132
Measurement Units		
Social measures		
Acres	3.14	7
Average daily traffic (ADT)	3.28	7
Other social measures [11 measures]	2.75	24
Physiographic measures [11 measures]	2.83	23
Biological measures [2 measures]	2.43	7
Economic measures		
Dollars	2.60	23
Other economic measures [8 measures]	2.53	15
Mixed-category, other [8 measures]	2.63	8
Mitigation measures [3 measures]	3.28	32
Verbal forecasts without clear measures	2.75	93
Certainty About Impact		
Quantified probability	4.00	1
Probability implied by "will" key words	2.97	133
Possibility implied by "may" key words	2.67	79
Other (conditional, ambiguous, etc.)	2.38	26
Average, full sample	2.816	239

Table 7.9
Predictive accuracy compared with forecasts' populations and timing. Means are averages of the accuracy ordinal ranks as shown in Table 7.2. Uncodable cases (e.g., unanticipated impacts) excluded from table.

Forecasts' Characteristics	$\overline{X}$ Accuracy	N
Location of Impacted Population or Resource		
Immediate project area	2.89	174
County, municipality	2.69	42
Economic sector	2.26	15
State or substate region	3.00	5
National	3.00	1
....		
Time Frame of Forecast Impact		
Construction or implementation period only	3.02	43
Short-term (i.e., relapsing to pre-project condition within a few years)	3.16	12
Long-term, permanent, indefinite	2.76	182
Average, full sample	2.816	239

The ideal EIS prediction of the prescriptive literature, as noted in Chapter 4, should specify the probability that the forecast impact will occur. The only forecast in the sample that conformed to this standard, the Paint Creek instream flow case, was correct. In the overwhelming majority of cases in the sample, readers must infer EIS writers' certainty about an impact based on the differences in key words in the forecast. As shown in Table 7.8, forecasts that convey high certainty by using key words like "will" are more accurate than those that imply only possibility by using key words like "could." This difference is partly attributable to the fact that—again—80% of the systematically accurate mitigation forecasts utilize high-certainty, "will" key words. In addition, one naturally finds a relatively high proportion of "impact within the range of a vague forecast" ratings among the cases whose forecasts used uncertain key words like "could" or "may." These cases thus also have a lower ratio of 4-point "close" ratings to 3-point "range . . . vague" ratings than the "will" key word forecasts. In fact, given these two factors, it is surprising that the accuracies of these two blocks of forecasts are so close.

Analyses of the populations impacted and time frames covered in the sample's forecasts are presented in Table 7.9. These characteristics are only indirectly related to forecasts' technical precision, as explained in Chapter 4, and analyses of them provide no additional insights into EIS forecast accuracy. Most forecasts deal with impacts in the immediate project area or in the community near the project. The accuracy difference between project-area and community impact forecasts and between short-term and long-term impacts are both artifacts of the accuracy of mitigation forecasts. That is, mitigations are usually carried out at or near the project site during the construction phase or in the short term after the project is carried out. Thus, the accuracy of construction-phase and short-term forecasts drops to 2.75 and 2.87, respectively, when mitigations are deleted from the averages. The low score in Table 7.9, the average of 2.26 for forecasts about economic sector effects, merely reflects the relative inaccuracy of economic forecasts.

Finally, we calculated the correlation between EISs' average accuracy and the forecast "vagueness" indices discussed in Chapter 4. Not surprisingly, since all the differences reviewed above have been small and mostly spurious, that correlation, $r = -.12$, is in the correct direction (decreased vagueness should be associated with increased accuracy), but is insignificant.

INSTITUTIONAL FACTORS

The political and organizational theories about the NEPA process are commonly divided into "internal reform" and "external reform" arguments. Briefly, the internal-reform position holds that NEPA forced agencies to recruit environmental specialists to prepare EISs, and that those staffers have struggled to increase resources agencies' consideration of environmental concerns (Fairfax and Andrews 1979, Taylor 1984). As agencies became more experienced with the NEPA

245

Table 7.10

Average accuracy ratings of the 29 EISs in the field sample. The accuracy means are averages of the ordinal ranks shown in Table 7.2.

EIS	$\bar{X}$ Accuracy	N
I-295 beltway, Florida (FHwA)	3.57	7
Florida highway 24 (FHwA)	3.33	6
Cummington radar facility (FAA)	3.25	8
Henrico wastewater treatment plant (EPA)	3.20	5
Oak Ridge Nat'l Lab. waste disposal (AEC)	3.14	7
South Fourche small watershed (SCS)	3.00	9
Jackson airport (NPS)	3.00	7
Hiwassee unit plan (FS)	3.00	6
BARC sewage treatment plants (ARS)	3.00	6
Cross-Florida barge canal restudy (Corps)	3.00	6
Sequoyah UF_6 plant (NRC)	2.90	11
Shepard Park development (HUD)	2.90	11
Newington Forest development (HUD)	2.90	10
Weymouth-Fore and Town Rivers (Corps)	2.83	6
Yakima CBD renewal (EDA)	2.81	11
Skippanon River bridge, Oregon (FHwA)	2.81	11
Cleveland Harbor disposal site (Corps)	2.80	5
Tacoma Harbor dredging (Corps)	2.77	9
Monticello nuclear plant (AEC)	2.75	12
Grand Teton N.P. master plan (NPS)	2.75	8
Paint Creek dam (Corps)	2.71	14
I-94 interchange, Michigan (FHwA)	2.71	7
Illinois Beach Park expansion (BOR)	2.62	8
Weyerhaeuser timber road (FS)	2.57	7
U.S. 53 and U.S. 12, Wisconsin (FHwA)	2.55	9
Alma unit No. 6 (REA)	2.54	11
Shippingport breeder reactor (ERDA)	2.50	8
Wisconsin highway 64 (FHwA)	2.40	5
Ozone unit plan (FS)	1.88	9
Average, whole sample	2.816	239

process, internal-reform proponents like Wichelman (1976) expected environmental values to be integrated into agency missions. The external-reform position holds that the NEPA process provided several avenues for public participation in previously closed agency decisionmaking, and that shift in interest group access enhanced environmentalist influence on resources decisions (Friesema and Culhane 1976). These arguments about the ways EIS writing affects agency decisionmaking are not mutually exclusive or antagonistic; for example, external environ-mentalist pressures can reinforce the position of EIS staffers within an agency.

Unfortunately, because internal-reform battles are hidden from public view in the recesses of intraagency decisionmaking, they are hard for outsiders to detect at the time an EIS process is underway—and impossible to document during a postaudit several years later.[6]   Thus, we have only the roughest indicators of political-institutional influences in EIS forecasting. Those indicators do not materially help explain forecast accuracy.

Individual EISs' accuracy provides a starting point for analyzing institutional factors because most of those factors (e.g., the mix of disciplines in an EIS-writing staff) are constants during the preparation of a given EIS. As shown in Table 7.10, the accuracy of the 29 EISs varies widely. The document with the best average accuracy rating, the I-295 FEIS, contains a very ordinary environmental assessment. Its high average is attributable to three mitigations rated "close" and an equal number of fuzzy cases that received index scores of "3," such as a long-range economic development forecast rated "not yet" and a residential and commercial displacement forecast rated "intuitively obvious accuracy"; its only creditable case is a "close"-rated traffic volume forecast. The Ozone unit plan brings up the rear, with an average a full 0.52 lower than the next-worst EIS. The plan and EIS were filed in 1976, just before Congress passed the National Forest Management Act, which mandated substantial changes in the Forest Service's land use planning process. The foresters on the Ozark National Forest then essentially put the plan on a shelf and forgot about it. The plan's nine audited impacts consist of one timber production forecast rated "close," one "impact within the range of a vague forecast," and seven inaccurate forecasts.

While there is substantial variance among individual EISs' accuracy, there—again—seem to be no significant patterns to the variance. As shown in Table 7.11, the average accuracy scores of five of the six lead agencies

Table 7.11
Predictive accuracy among project types and lead agencies with more than one EIS in the sample. Means are averages of the accuracy ordinal ranks as shown in Table 7.2.

EIS Subsamples	X̄ Accuracy	N
Lead Agency		
Nuclear Regulatory Commission (+ AEC)	2.90	30
Dept. Housing and Urban Development	2.90	21
Federal Highway Administration	2.88	45
National Park Service	2.86	15
Corps of Engineers	2.80	40
Forest Service	2.40	22
Project Type		
Housing, urban, buildings	2.98	58
Highways	2.88	45
Water resources	2.83	49
Energy (nuclear and coal)	2.76	49
Public lands	2.52	38
Average, full sample	2.816	239

with two or more EISs represented in the sample are tightly bunched between 2.9 and 2.8. (The scores of the other seven agencies represented in the sample are simply their EISs' averages, as listed in Table 7.10.) The Forest Service has a significantly lower 2.4 average, but if its Ozone plan millstone had not been in the sample it would have had a respectable 2.77 average. The Forest Service and Corps of Engineers have been widely credited with significant improvements under NEPA (cf. Mazmanian and Nienaber 1979, Culhane 1981, Taylor 1984), and they may have indeed implemented good NEPA policies in the face of sustained environmentalist criticisms, but their averages in the sample do not demonstrate any superiority at environmental forecasting. On the other hand, the NRC, whose NEPA

248

procedures have been attacked in such precedential litigation as the Calvert Cliffs (1971) and Vermont Yankee (1978) cases, is tied for first place among the agencies represented in the sample.

There are also few notable differences in forecast accuracy among the five general types of projects. Public lands EISs have the worst accuracy average, which is again attributable to the Ozone EIS. Land use plans might seem prone to inaccuracy and imprecision since they deal with a wide range of types of agency actions, spread over moderately large geographic planning areas, and implemented over time frames of a decade or more. However, the other two public lands use plans averaged accuracy ratings at and above the sample mean. Thus, the Ozone plan seems to be simply an unusually bad one. If we ignore the deviant Ozone plan, the remaining public lands EISs average 2.78, which is within the highway, water resources, and energy EISs' range of only +0.06 from the sample mean.

The housing and buildings EISs, which really constitute a catchall project category, had the highest average accuracy for several reasons. First, the HUD EISs in the sample are very competently written documents with above-average accuracy. Second, the government building EISs contain more than their share of "3" rankings; for example, the Cummington radar cases include four "impact within the range of a vague forecast" and only one nontrivial "close" rating. Third, over half of the housing and buildings cases are social forecast impacts, which are generally more accurate than average.

We found no evidence in our sample of any systematic improvement in forecasting over time. That is, there is no correlation ($r = -.02$) between FEISs' filing dates and their average accuracy, and only an insignificant correlation ($r = .12$) between projects' implementation dates and their EISs' accuracy. Of course, our EISs were written during a narrow time period in NEPA's history—after the initial shakedown years of the early 1970s and before the implementation of CEQ's (1978) regulations. We know there is a notable difference between the EIS prepared during NEPA's first few years and those written later, and we also believe there is little or no essential difference between the EISs written before and after 1978. However, our data cannot confirm or refute those beliefs.

Finally, we found no evidence that forecast accuracy is related to external-reform influences. Our usual indicator of the amount of diffuse external, "peer review" pressure on lead agency EIS writers is the number of comment letters

received on a DEIS.[8] Not all letters are equal, of
course, since letters from EPA, the departments of Interior
and Commerce, and some other agencies carry greater statutory and professional weight than a handwritten note from an
ordinary citizen. Nonetheless, the total number of comment
letters provides a good, simple quantitative indicator of
public pressure on an EIS. This measure is not at all correlated (r = -.05) with EISs' forecast accuracy.

Nor is an EIS's controversiality related to its accuracy. As noted in Chapters 3 and 4, four EISs in the field
sample dealt with some significant national controversy:
the Cross-Florida barge canal, an infamous project involving
litigation, a CEQ referral, and presidential decisions; the
Jackson airport, which also involved several rounds of
litigation, a CEQ referral, secretarial-level decisions,
plus multiple EISs; the Monticello plant, which was the
subject of a leading precedent on federal preemption of
state regulatory authority in the nuclear licensing field;
and the Shippingport breeder reactor, whose multivolume EIS
generated a lengthy, national, polarized comment record.
These four EISs' have an average accuracy of 2.79. The five
EISs involving litigation--Cross-Florida, Jackson, and
Monticello, plus two local-litigation projects, the U.S. 53
highway project and the Cummington radar building--have an
accuracy average of 2.88. All six controversial EISs (i.e.,
the litigation EISs plus Shippingport) average a 2.82
accuracy rating, that is, dead on the sample mean.

CASE-BY-CASE EXPLANATIONS

Since none of the standard theories about EISs account
for forecasts' accuracy, we now turn to ad hoc explanations.
The final code assigned to a case is a brief comment that
summarizes any apparent reason for the case's accuracy
rating. These codes do not provide us with explanations
that are much more helpful than those reviewed thus far.

On the one hand, this open-ended coding produced 55
distinct explanations about the 239 cases in the sample.
Fully 25 of these explanations apply to only a single case
in the sample. Several explanations are quite interesting
and are reviewed in Chapter 6. For example, in the Cleveland harbor power plant case, the Corps ignored an apt EPA
comment, and the Jackson airport's passenger loads seemed
to increase because of its new electronic landing equipment,
not the jet service discussed in the EIS. The sample
includes more interesting cases that, in the interest of

Table 7.12
Auditor's characterizations of primary reasons for forecast accuracy ratings. Only explanations with four or more cases are shown below. A total of 55 distinct explanation codes were assigned among all 239 cases. Accuracy means are averages of the accuracy ranks as shown in Table 7.1.

Explanations	N	$\bar{X}$ Accuracy
For Relatively Accurate Forecasts		
Presumptive administrative effectiveness (by EIS writers or impl. officials)	53	3.43
Incidental, commonplace impact	21	3.09
Standard operating procedure of agency or project sponsor, known	13	3.62
Effective implementation (e.g., of mitigation), known	8	3.75
Initiative by outside/nonagency actors	7	3.71
Nonfalsifiable, very vague, or "puff" forecasts	7	3.00
Intervention of political actors	5	3.20
Postdictions	4	3.50
For Relatively Inaccurate Forecasts		
Exogenous, unpredictable change	17	2.18
Contingent event did not occur	13	2.38
Poor EIS forecasting, known or apparent	12	1.83
Too soon for impact to occur (including time-random events)	9	2.67
Conservative forecast	6	2.33
Implementation error, problem, or delay (known)	5	2.00
Impact dwarfed by large adjacent project	4	2.50
Impact affects small proportion of (measured) population	4	2.50
Equipment failure	4	2.00

conserving space, are not reviewed in Chapter 6. In any event, the reader should understand that there are a large number of unique apparent explanations for many forecasts' accuracy.

On the other hand, the ad hoc explanations that recur frequently do not tell us much about good environmental forecasting. The most common reason for an accurate forecast is "presumptive administrative effectiveness" (Table 7.12). This simply means that our audit turned up no apparent reason for the case's accuracy other than the presumption that the EIS writers made a satisfactory forecast about project consequences or the agency's implementing officers effectively managed the project to bring about the forecast outcome. In about a dozen cases, a correctly forecast outcome could be attributed to routine organizational procedures of the lead agency or project sponsor, and in eight cases we knew about nonroutine implementation actions (e.g., the special Weymouth-Fore blasting procedures and monitoring) that had ensured a forecast outcome.

Some other common explanations for accurate cases indicate that the relevant forecasts are rather lame. The second most common reason for an accurate forecast is that the impact was either incidental or very commonplace. Another dozen forecasts proved to be accurate because they were essentially nonfalsifiable, trivially obvious, or statements about impacts (such as Sequoyah and Monticello property tax payments, described in Chapter 6) that had occurred by the time the EIS was written. Seven outcomes turned out as forecast chiefly because outside professionals (such as the university scientists who performed the Paint Creek archeology survey and fishery study), rather than the lead agency's responsible officers, arranged for the studies described in the FEIS. In the bottom half of Table 7.12, another half-dozen forecasts were rated inaccurate because they were manifestly conservative, if not overly conservative. None of these cases reflect much credit on the EIS writers involved.

Among the more interesting reasons listed in Table 7.12, 30 cases were rated inaccurate essentially because EIS writers made incorrect ceteris paribus assumptions. The "exogenous change" explanation covers a wide variety of occurrences of some change unrelated to the project, which EIS writers can hardly be faulted for not predicting. These cases include the power plant closing in the Weymouth-Fore shipping case, the new Arkansas loose cattle law in the Ozone grazing case, and the statutory formula change in the Hiwassee PILT case. In almost as many cases, EIS writers had forecast an impact depended on some intervening event, and when the event did not occur, neither did the impact.

We also note that several interesting aspects of environmental management that we had thought important

(Friesema and Culhane 1976, Culhane and Friesema 1977) seemed to be determining factors in only a few cases. Only four forecasts involved impacts that were overwhelmed by spillover effects from contemporaneously completed, nearby projects. Three of these forecasts were made in the EIS on the Shippingport breeder reactor, a 150 mWe research unit installed in the mid-1970s at the same time that a huge 850 mWe nuclear station was built next door and an equally large coal-fired station was built on the opposite side of Shippingport, a tiny town of 255 souls. In addition, while at least six projects were significantly affected by complex politics involving the courts or high-level executive branch officials, the accuracy of only five forecast impacts per se seemed to be determined by those politics. In other words, in searching for reasons to explain why some forecasts are better than others, we have come up dry once again.

CONCLUSIONS

We found two basic patterns of accuracy among the 239 forecast impacts in the field sample. First, EIS forecasts are not inaccurate. We use this double-negative to highlight our finding that very few impacts in the sample are demonstrably inconsistent with EIS forecasts. Even fewer impacts are unanticipated, and no egregious unanticipated impacts could be identified during the study. On the other hand, only about a third of the forecasts in the study are particularly accurate. The more numerous, middling forecast impacts are either accurate solely by virtue of the vagueness of the forecast or somewhat inaccurate in various complicated ways.

The logic of environmental assessment in the prescriptive literature on NEPA demands accurate EIS forecasts. A rational analyst must weigh a comprehensive range of consequences of each alternative course of action. This analyst's predicted consequences must be accurate, for otherwise the decision resulting from his or her weighing of those consequences cannot be assumed to be optimum. Impacts rated "within range of a vague forecast" and other "grey area" cases clearly fall short of the rationalist ideal of environmental assessment. That is, a forecast that is either unquantifiably vague or quantitatively wrong is not very useful in calculating a project's net benefits and choosing an optimum alternative.

Second, in our proposal for this project, we suspected that we might find "unsystematic significant variance"

253

between actual impacts and forecasts (Culhane and Friesema 1981: 8). The variance in forecast accuracy is certainly significant, but we did not believe it would be so unsystematic. Mitigation promises are generally quite accurate. The fact that these actions are generally carried out reinforces policy theorists' notions that the complexity of joint action hampers project implementation (cf. Pressman and Wildavsky 1973, Culhane and Friesema 1977, and Chapter 3 above). That is, mitigations involve fairly well defined actions that are usually within the control of lead agency managers. Despite some general cynicism about the veracity of government promises, agency managers prove to be quite responsible in carrying out promised mitigations. Beyond the systematic influence of accurate mitigations in the analyses discussed above, there are virtually no noteworthy patterns, factors, or variables that explain either accurate or inaccurate forecasts.

Rather than overanalyzing obviously insignificant 0.1 differences of means, let us instead focus on some core truths. EIS writers are not cynically deceptive, nor are they presciently rational-comprehensive analysts. They are simply humans who are subject to all the analytical limitations and inclinations towards satisficing decision-making that Nobel laureate Herbert Simon (1947) first described four decades ago.

NOTES

1. For this reason, while we have excellent data on sediment toxicity in Tacoma harbor, we rated the quality of our data on this case as "low." That is, we knew some impact had to have occurred, but we cannot document the magnitude of that impact.

2. For example, in the Weymouth-Fore shipping efficiency case reviewed in Chapter 6, shallower-draft ships actually used the harbor following the Corps' channel deepening project. Since ship drafts are the critical indicator of the project's economic benefits, this wrong-direction outcome received a primary rating of "incon-sistent." However, the apparent cause of this inconsistent outcome was the closure of Weymouth-Fore's primary deep-draft wharf because of the so-called energy crisis of the 1970s. Thus, the case also received an exculpatory sec-

ondary rating indicating that the impact was "spurious."

3. Even though we believe that adversarial review of EISs is the NEPA process's key reform of resources decision-making, we did not subscribe to the cynical benefit-overestimation, cost-underestimation hypothesis. For one thing, public and interagency review of EISs serves to keep EIS writers honest, just as does peer review in normal science. Nonetheless, our NSF proposal treated the benefit-overestimation, cost-underestimation charge as a nominal hypothesis, while expecting the hypothesis to be wrong (Culhane and Friesema 1982: 7-11). For a review of the early NEPA literature supporting this cynical view, see Cortner (1976: 324). However, the best "citation" for this view would be the typical militant environmental group's comment on any controversial EIS.

4. Even though our fieldwork protocol called for an audit of each EIS's most controversial impact, several EISs did not contain any truly controversial forecasts. The 21 controversial cases include 18 impacts, two objectives, and one mitigation.

5. See, for example, Beanlands and Duinker's (1983) framework. By technical precision, we mean quantification and the other characteristics of an ideal prediction, as defined in Chapter 4. The indicators of forecast precision used through this book conform to the spirit of the indicators of EISs' technical quality discussed in our NSF proposal (Culhane and Friesema 1982: 8-11; also cf. Friesema 1981). That is, forecast quantification involves the same style of forecasting implied by the proposal's "collection of site-specific data," "use of models," and so forth. The analysis reported above may be superior to the one proposed in 1982 because we can compare a forecast's accuracy ot its precision, rather than to EIS-wide factors that may be irrelevant to the forecast.

6. The averages mentioned in this paragraph all exclude mitigations, since mitigations are coded as a special nominal measure. Of course, the accuracy of the mitigation forecasts is the same in Tables 7.3 and 7.8.

7. Certain information, which we proposed to acquire as indicators of institutional influences on the EIS process (Culhane and Friesema 1981: 11), proved unavailable. In particular, information on the composition of EIS-writing teams, which has been routinely printed in FEISs since CEQ's (1978) regulations, is missing in most of the sample EISs and is as difficult to glean from agency records as are postproject impact data. Also see note 8 below.

8
Systems Analysis or Simplicity?

Resources policymaking has changed substantially since the passage of the National Environmental Policy Act. NEPA's first decade was accompanied by a diffuse shift from public approbation for "pork barrel" projects that promised local economic benefits to public suspicion about the environmental, social, and fiscal costs of such projects. The numbers of projects proposed by the major resources development agencies certainly declined during the period covered by this study. For example, the numbers of EISs filed by the major water project agencies (the Corps of Engineers, Bureau of Reclamation, and Soil Conservation Service) fell from an average of 305 EISs per year during 1971-1976 to 129 per year during 1977-1982.[1] Similar patterns can be seen in project proposals by agencies as diverse as the Forest Service and the FHwA. We do not know whether this shift was a product of NEPA per se or the politics of the "environmental decade" generally. NEPA, however, certainly gave those political pressures a legitimate channel into the bureaucratic world of project decisionmaking.

This study, however, has not dealt with the macropolitical or macropolicy impacts of NEPA. A principal concern of this book has been whether EISs embody good rational-comprehensive analyses. Specifically, we first examined whether EISs contain the kind of quantified, precise forecasts that are the hallmark of state-of-the-art environmental assessment, according to the prescriptive literature. Second, our search for data on project impacts led us to certain conclusions about whether resource project managers act like the self-evaluating officials envisioned by the scientific-rational model. Finally, we asked our bottom-line question: If EISs are to assist in selecting comprehesively optimal projects, then ought not their forecasts be correct?

The rational model we have described is not principally the progeny of NEPA's congressional sponsors. They and their staffers may have succumbed to the allure of PPB, cost-benefit analysis, and similar reforms popular in the 1960s, and thus sprinkled rational-comprehensive and scientific language throughout the act. But we cannot picture top-drawer politicians like Senators Edmund Muskie and Henry Jackson, Representative John Dingell, and congressional staffer Daniel Dreyfus as proponents of full-blown systems analysis. Actually, federal agencies' lawyers began to insist that EISs mention anything and everything so they would not have to litigate their failure to consider some impact. The rational model of environmental assessment is more a product of EISs' enthusiastic supporters within the scientific and technical community, such as Keith Caldwell, who played a major role in NEPA's drafting, plus Luna Leopold et al. (1971), Derek Medford (1973), Larry Canter (1977), Larry Leistritz and Steve Murdock (1981), Gordon Beanlands and Peter Duinker (1983), Bob Bartlett (1986), and others. That model also, in effect, provided a justification for the increasingly comprehensive, or at least lengthy, documents EISs had become.

Having been trained and socialized in the scientific method, we too admire state-of-the-art environmental assessment methodologies and sympathize with the vision of the "S&T" enthusiasts. Nonetheless, at this point we must ask, if the rational-scientific model does not aptly describe EIS writing, what should we think about these documents?

RATIONAL ANALYSIS AND EISs

This study's empirical case against the rational model of EIS writing is based on four principal findings about the sample EISs.[2] First, most of the 1,105 forecasts in the EISs are much more imprecise than the ideal prediction of the prescriptive literature. The average forecast in the sample is unquantified, is not couched in any recognizable unit of measurement, and lacks any clear statement of the significance or likelihood of the impact. Only 14% of the sample forecasts achieved a perfect score on our index, described in Chapter 4, which counts the number of imprecise characteristics in a forecast.

Second, EIS proposals are not implemented in a straight-forward, orderly fashion. As noted in Chapter 3, EIS proposals are subject to significant buffeting during the project planning and implementation process. Technical, fiscal, legal, and other implementation pathologies infected two fifths of the implementation sample of 146 EISs, causing abnormal delays, suspensions, or cancellations in 18% of the projects. Thus, project implementation is a controversial political, not rational-comprehensive, planning process.

Third, agency managers do not systematically evaluate their projects to ensure that the EIS's predicted net-benefit mix actually occurs. In seeking data on project impacts, we were able to locate good quantified data for only a third of the impacts selected for field audits,[3] and many of these data were not in the possession of project managers. Of course, with a few exceptions, project managers are not legally bound to monitor their projects. In any case, after struggling to get projects up and running, it appears that lead agencies relax and pay only unsystematic attention to the projects' consequences.

Fourth, only a minority of the forecasts audited proved tolerably accurate. As described in Chapters 6 and 7, only 30% of the forecasts received accurate audit classifications ("close" or "complex, but accurate"). On the other hand, 27% of the impacts were not inaccurate solely by virtue of their forecasts' vagueness. Another quarter of the cases received "grey area" classifications indicating the forecast was quantitatively or qualitatively inaccurate, if not wrong enough to receive an "inconsistent" classification.

Our findings also appear consistent with the small body of international evaluations of environmental assessments. For example, the so-called Dutch study found that, in a sample of 140 EISs from western Europe, the British Commonwealth, and the U.S., only a quarter of the predictions used "quantitative techniques" (Environmental Resources Ltd. 1984). In an earlier case study, we found that power-plant EISs reflected state-of-the-art scientific understandings about acid deposition impacts in, again, only a quarter of the eligible cases (Culhane, Armentano, and Friesema 1985). Rosenberg et al. (1981) also found the predictions in a sample of what the authors regarded as the best Canadian and U.S. environmental assessments to be below "average" on their evaluation scale.[4]

Two other studies have examined both the predictive accuracy and technical precision of EIS forecasts. First, the University of Aberdeen group audited four British environmental statements. Bisset (1984: 468) noted that these documents' "predictions were . . . often expressed in vague, imprecise, and woolly language." The Aberdeen group found that only 39 of its 100 auditable predictions were clearly accurate. Half that number of predictions (19) were deleted from the audit solely because they were too vague to audit. In addition, they found 19 predicted impacts that did not occur, four predictions of no impact in which some impact occurred, nine underestimates, seven overestimates, and five predicted impacts that had not yet occurred (Bisset 1984: 470-471). Both the range of accuracy classifications the Aberdeen group used and their distribution of predictive accuracy are quite comparable to our findings.

Second, Leistritz and Murdock, two authors in the state-of-the-art prescriptive literature on EISs discussed in earlier chapters, evaluated population predictions in 225 U.S. EISs. Murdock et al. (1982: 336) noted a "striking absence" of adequate quantified forecasts and baseline information about population changes. In fact, almost half contained no meaningful population forecasts, which Murdock and Leistritz regard as an essential basis for any competent socioeconomic impact assessment. Among the 20% of EISs that contained auditable population predictions, they found a mean absolute percent error of 15.4% for county-level 1970-1980 population-change predictions and 54.7% for city-level predictions.[5] Since population forecasting is one of the most straightforward types of socioeconomic analysis, the authors regarded these error rates as very inaccurate.

These studies methods vary considerably. Researchers who select one impact type, such as population (Murdock et al. 1982) or acid deposition (Culhane, Armentano, and Friesema 1985), can conduct a methodologically more pleasing evaluation. Researchers who chose to examine the whole range of impacts covered in EISs, such as Rosenberg et al. (1981), Bisset (1984), and we, must make do with less satisfying classifications. But the bottom-line conclusion of these studies, irrespective of methodology or country in which the EIS is written, is the same. EIS forecasts are usually not technically precise or particularly accurate.

A substantial literature in organization theory explains why decisionmaking differs radically from the assumptions of the rational model. Relying on human intuition alone, decisionmakers cannot easily identify all alternatives and consequences or calculate the optimum solution in the matrix of alternatives and consequences implicit in even a fairly simple EIS. Really comprehensive analysis, as in Quade's (1964) systems analysis, is quite expensive. Thus, unless forced to engage in such analysis, almost all organizations find that the marginal costs of systems analysis greatly exceed the marginal benefits of an optimum, as opposed to satisficing, decision (March and Simon 1958). Most important, as Lindblom (1959) and his followers note, in political decisions--and the NEPA process is a political decision process--participants can rarely come to agreement about the goals or preference functions that are assumed to be settled before rational-comprehensive analysis begins.

The NEPA process, as elaborated by the prescriptive literature based on many cues in the act's language, nonetheless holds out rational-comprehensive decisionmaking as an ideal. In addition, the prospects for conducting comprehensive EIS analyses are better than for doing PPB or ZBB budgeting or cost-benefit analyses of environmental or safety regulations. The latter place the analyst in the vortex of visible, national policy controversies. EISs, by contrast, are usually prepared by agency technical staffers (or consultants) responsible to low-level line officers. Administrative theory tells us these people should be insulated by bureaucratic routines from political influences, and thus more capable of neutral rational-technical analysis than, for example, congressional and OMB budgeters. (On the other hand, theory aside, we also know local natural resources managers face active political constituencies; cf. Mazmanian and Nienaber 1979, Culhane 1981.)

This study's field sample consists of 29 U.S. EISs written during the mid-1970s on projects generally implemented during the late 1970s. Environmental analyses may someday come closer to the rational-comprehensive ideal. Certainly the scientific community had not provided the environmental analysts of the 1970s with all the scientific knowledge and tools they needed in some key areas. We understand that some of the assessments currently being conducted on so-called Superfund decisions come much closer to the model of quantitative, comprehensive, state-of-the-art analysis.[6] (Of course, rational-comprehensive analysis is appropriate for high-cost, high-risk Superfund decisions. Such rational decisionmaking is also appropriate for nuclear regulatory decisions, and we found that AEC/NRC EISs came generally closer to the model in our field sample.) Moreover, the Forest Service has begun a systematic program to monitoring of a broad range of impacts and objectives after implementation of its land management plans.[7]

Nonetheless, our field-sample EISs fall far short of the ideal of technically rational, comprehensive, optimizing analysis.

261

ENVIRONMENTAL IMPACTS AND EISs

Testing whether real-world governmental officials' act like rational-comprehensive analysts seems about as fair as challenging your grandmother to an arm-wrestling contest. However, we ought to expect certain things from environmental impact statements. By common understanding, EISs should examine the trade-offs between impacts on the natural environment and economic or other objectives, and that interpretation has been incorporated into NEPA legal precedents, such as the 1971 Calvert Cliffs opinion's treatment of "balancing" decisionmaking. Our findings suggest two empirical caveats about this common understanding.

On the one hand, EISs devote substantial space to discussions of social impacts. As noted in Chapter 2, whatever the abstract meaning of NEPA's "affecting the quality of the human environment" clause, impacts on the humans in a project's environs determine the net public support for or opposition to a proposal. Social impacts thus significantly affect the amount of conflict agency officers must face during public and interagency review of an EIS. Formal NEPA law, however, assigns social impacts a second-class status. Following the 1978 Image of Greater San Antonio decision, the CEQ (1978: §1508.6) regulations stated that socioeconomic impacts alone are insufficient to trigger NEPA's EIS requirement. Once that requirement is triggered, however, the EIS should address social impacts as thoroughly as other impacts. The Supreme Court's 1983 PANE decision, which held the asserted psychological impact of neighbors' fears about the restart of the undamaged Three Mile Island reactor did not require an EIS, arguably confirmed CEQ's (1978) stance.

Nonetheless, social impacts occupy no second-class niche in our sample of EISs. They account for 40% of the forecasts in the field-sample EISs, the largest percentage among the four substantive impact categories in our analysis. Social data proved to be the most available of the four categories during our fieldwork, and those data are reasonably well quantified (though not so quantified as economic data). Moreover, social forecast impacts had the highest average accuracy classifications among the four categories. Thus, legal nuances debated late in the 1970s notwithstanding, EISs written by this time had come to focus disproportionately on impacts on people.

On the other hand, EISs' biological impact assessments seem woefully underdeveloped. We found fewer forecasts of biological impacts than of other categories. Biological forecasts were the least quantified, rarely used available biological concepts and measures such as the Shannon species diversity index, and had the highest vagueness index scores in the sample. In auditing forecasts, it was difficult to find satisfactory biological impact data. The biological impact information we found was generally unquantified (only 36% quantified, versus 57% for the rest of field data), and had the worst net data adequacy rating among the four categories. The paucity of biological forecasts and impact data allowed us to audit only half as many biological forecasts as we audited in the other categories, though those few forecasts were about as accurate as the sample average.

One could argue that NEPA was directed more at impacts on the biological components of natural environments than at physiographic impacts. In 1969 the most often-mentioned physiographic impacts—air and water pollution and radiological effects—were regulated by preexisting statutes administered by established federal agencies (though some of those agencies were reorganized in 1970 to form EPA, and the air and water pollution statutes were tightened up in 1970 and 1972). In any case, we found little or none of the kind of holistic evaluation of ecosystem impacts that would be responsive to NEPA's mandate to "prevent or eliminate damage to the environment and biosphere," "enrich the understanding of ecological systems," "utilize ecological information in the planning and development of resource-oriented projects," and so forth.[8]

We do not advocate any dimunition of social impact assessment. Even if we did, the public review of EISs would continue to lead EIS writers to be sensitive to impacts on the local people who read EISs. However, we do believe that EISs' biological discussions ought to be better informed and more sophisticated than most of the material we examined in our field-sample documents.

ENVIRONMENTAL AUDITING

When we embarked on this project, we hoped that it might document the utility of environmental auditing as a capstone of the NEPA process. There are at least three views of useful roles EISs can play in project management. First, an EIS can be viewed as an implied contract with a lead agency's overseers and constituencies about the forecast mix of natural and socioeconomic environmental benefits and adverse impacts (cf. Environmental Law Institute 1980). That is, according to this view, project opponents and prospective plaintiffs can profit more by auditing actual project consequences and holding an agency to its implied promises than by picking at the nits in draft EIS phrasing. Second, consultants recommended environmental auditing to their business clients as a defensive measure to ensure corporate compliance with environmental regulations (cf. A. D. Little 1981, Reed 1983, Greeno 1983, and Harrison 1984). Third, environmental postauditing logically complements a management-by-objectives (MBO) strategy of project administration (cf. Drucker 1976). That is, the EIS's list of objectives, necessary mitigations, and critical forecast impacts constitutes a set of criteria to guide managers who wish to ensure that their projects are effective.

We discovered one datum supporting the MBO theory of post-EIS administration during our fieldwork. Guy Thurmond, district ranger on the Hiwassee District, described the Hiwassee unit plan as "sort of like a Bible" and enthused, "I get a heck of a lot of management direction out of it." He said his district had "tiered off" the plan; that is, his staff took the plan's management directions and translated them into detailed prescriptions for the management compartments within the planning unit.[9] In particular, in a classic Forest Service political-managerial observation (cf. Kaufman 1960), Thurmond noted that the "plan solves a lot of bickering" within and among the district and forest supervisor's office staffs and the district's external constituencies. That is, a new line officer like Thurmond, who was transferred onto the district three years after the plan's approval, could use the plan to prevent a rearguing of issues among protagonists dissatisfied with prior accommodations and to control clients. Ranger Thurmond, who had served as a planner on the staff of another national forest before becoming Hiwassee district ranger, was clearly a NEPA aficionado. Nonetheless, he was the only manager in 29 projects who appreciated an EIS as a management tool, rather than as predecision paperwork.

Management-oriented environmental auditing fits easiest into a rational-scientific view of project planning and implementation. For example, the Beanlands-Duinker framework (Figure 5.2) views audits as the last step in a comprehensive systems-analysis process in which an EIS's hypotheses about project impacts are tested. Notwithstanding some of Ranger Thurmond's sophisiticated observations and the existence of some regulation-averse corporate environmental audit programs, our results make us less than sangiune about the prospects for routine postauditing as a capstone to the NEPA process. As noted in Chapter 5, most lead agencies show little or no interest in their projects' actual consequences. Max Weber, the German sociologist, noted that one of the principal characteristics of a bureaucracy was that it keeps its important information in the files. Since we could find so little systematic data on forecast impacts in lead agencies' files, we conclude that information was not important to those bureaucracies. Rather, the conventional wisdom suggests capital project and natural resources agencies reward a manager for planning a project, building a coalition supporting decisions favoring the project, and implementing that project. What happens thereafter is immaterial, since the manager is building her next project. More generally, if EIS writing does not conform to the rational decision model, how could one expect such an ideal response to some new rational-comprehensive postproject stage that might be added to the NEPA process?

RETHINKING EISs

The EISs we audited clearly fail the relevant tests of good rational-technical analyses. However, we hope our evaluation will not be read as an attack on the NEPA process or EISs per se. The flaw lies as much with the version of the rational-comprehensive model espoused by the prescriptive literature as with the shortcomings of real-world EISs.

Many EIS passages that fail the test of state-of-the-art ideal prediction are, nonetheless, quite sensible assessments that would pass a "reasonable person" test. That is, many of our sample's forecasts with poor "vagueness index" scores, or ones that received suboptimal accuracy classifications like "impact within the range of a vague forecast" or "impact less than forecast," could appear quite satisfactory under this test. Two forecasts encountered during the study demonstrate the differences between a good quantitative-form prediction and a satisfactory commonsense assessment.

On the one hand, Chapter 6 presented the Shepard Park development's school enrollment discussion as an example of the problems posed in auditing a vague forecast. HUD's EIS writers noted that, becase of almost a decade of declining enrollments, "the [St. Paul] School district has sufficient capacity to handle the additional students from the Shepard Park development." This passage is untestable by any standard of quantitative analysis. It does not even forecast the direction—increase, no change, or decrease—of postproject enrollment change. Yet the passage is quite satisfactory by a commonsense standard. Large apartment building developments can place a heavy burden on local school systems, so HUD's standard EIS impact checklist called for an assessment of school enrollment impacts. However, Shepard Park contains a senior citizens building, whose subsidy was the EIS-triggering action, plus a half-dozen luxury high- and mid-rise buildings. These buildings would be occupied by many fewer children, particularly primary-school-age children, than a standard apartment complex. In addition, HUD staffers knew that enrollment at the local primary school had declined by almost half during the 1970s. The few Shepard Park children would cause no more than a blip on the enrollment decline charted in Figure 6.18. The EIS's statement on the matter was, therefore, sensible and appropriate under the circumstances.

On the other hand, Chapter 4 presented HUD's Clovis Heights EIS's air-quality section as an example of an ideal assessment by quantitative and technical-format standards, but as one whose conclusion patently offends common sense. That section quantitatively modeled automobile and heating furnace emissions into a prediction of ambient air quality that would be worse than regulatory standards. However, the best that can be said for that section's concluding recommended mitigation--that homeowners turn on their air conditioners and close their windows tightly to keep pollutants out of the house--is that these EIS writers must have been putting their readers on.

These, of course, are deviant cases. Some imprecise verbal discussions are appropriate in light of the insignificance of an impact or EIS writers' legitimate uncertainties about an impact. Other verbal discussions are sufficiently precise that a quantitatively attuned reader could translate the passage into a quantified forecast. But many more EIS passages are unquantified, unquantifiable, vague verbiage that reflects sloppy environmental assessment. Some other EIS passages are overquantified. We previously commented on the imbalance between powerplant EISs' elaborately quantified predictions of pollutant dispersion in the plants' immediate neighborhoods and their insensitivity to, if not disregard for, acid deposition impacts (Culhane, Armentano, and Friesema 1985: 375). Nonetheless, a concise, quantified impact prediction is usually preferable, everything else being equal.

Both natural scientists and social scientists have been trained to understand that the most competent way to predict the future is to quantify based on validated theories and models, holistically weigh all relevant factors, and chose the best course of action identified by that analysis. That is, someone trained in the scientific method cannot feel instinctively comfortable with a method of decisionmaking explicitly defined as suboptimizing.

On the other hand, rational quantitative analysis can be performed best--perhaps only--within the theoretical consensus of a homogeneous discipline or cohesive profession. As environmental scientists, we can agree that a prescient, calculated analysis that weighs all impacts within a state-of-the-art scientific framework is best. We can agree that EISs ought to consider secondary impacts along with obvious primary impacts, regional or "mesoscale" impacts together with mundane local impacts, sophisticated biological impacts along with ordinary physical or socio-economic impacts, and so forth.

However, as organization theorists, we also realize how hard it is for GS-11 and GS-12 staffers to be so scientifically sophisticated under tight project deadlines, working in multidisciplinary teams where comparably credentialed professionals do not always see eye to eye, and enveloped in an adversarial political process in which all sides believe the bureaucrats have just not gotten things right. That is, one cannot conduct real rational-comprehensive analysis or implement proposals in a straightforward way without some solid institutional agreement about decision goals and premises.

A better model than the rationalist systems analysis of the prescriptive literature on EISs can be found, we believe, in CEQ's 1978 NEPA regulations. As the preamble to those regulations put it, "[I]t is not better documents but better decisions that count. NEPA's purpose is not to generate paperwork--even excellent paperwork--but to foster excellent action" (CEQ 1978: §1500.1). CEQ envisioned a good EIS to be not an elaborate technical document, but an aid in a decision process in which reasonable people were guided by relevant information and common sense. It per-ceived that EISs had grown to be lengthy documents and, worse, much of the padding in them was added by agencies' lawyers to mislead judges and other reviewing decisionmakers into believing comprehensive technical analyses had been performed. CEQ wanted shorter and better EISs.

CEQ (1978) stressed three key objectives in its regulations' initial "purpose and policy" section, various specific provisions, and its explanatory summary in the Federal Register.[10] First, EISs should be readable. Section 1500.2(b), the regulations key general statement of policy regarding EISs, states, "Environmental impact statements shall be concise, clear, and to the point. . . ." Section 1500.4(d) required that EISs be written "in plain language." These documents should be readable because, again according to the general policy of section 1500.2(b), they should be "useful to decisionmakers and the public."

Second, EISs should be shorter. CEQ (1978: §1500.4) led its list of reforms with the following injunctions:

Agencies shall reduce excessive paperwork by:
(a) Reducing the length of environmental impact statements by . . . setting appropriate page limits.
(b) Preparing analytic rather than encyclopedic environmental impact statements.
(c) Discussing only briefly issues other than significant ones.

The regulation's most specific reform in pursuit of parsimony was to prescribe page limits: the text sections of a final EIS "shall normally be less than 150 pages and for projects of unusual scope or complexity shall normally be less than 300 pages" (§1502.7). The regulations define "text" to include the project description, affected environment, alternatives, and environmental consequences sections; that is, only the front summary, public review and comment section, and appendices are excluded from the page limit. However, many other provisions, such as the authorization to cite or "incorporate by reference" lengthy but publicly available studies or documents (§1500.4(j), §1502.21), are also intended to shorten EISs.

Third, CEQ believed that acute analysis need not be sacrificed in shortening and clarifying EISs. The link between reducing verbiage while maintaining, indeed improving, analysis is repeated again and again in the regulations. The regulations' key statement of policy, (§1500.2(b), partially quoted above) states:

Agencies shall . . . reduce paperwork and the accumulation of extraneous background data and . . . emphasize real environmental issues and alternatives. Environmental impact statements shall be concise, clear, and to the point, and shall be supported by evidence that agencies have made the necessary environmental analyses.

Sections 1500.4(b) and 1500.4(c), quoted above, also assert such a link. In other words, agency staffers could be expected to conduct perceptive analysis--even state-of-the-

268

art quantitative analysis--but they must first stop stuffing EISs with reams of nonanalysis.

The regulations envisioned that, by cutting out extraneous and encyclopedic data, EIS writers would be forced to examine important but generally unexamined issues. CEQ wanted agencies, in particular, to assess reasonable, real alternatives to the proposed action (§1500.2(e)), rather than the bogus alternatives that agencies regularly used to make the proposed action appear to be the only viable decision. In addition, the regulations require discussion of both direct and indirect impacts (§1502.16, §1508.8) and imply that both should receive comparable attention. In this context, the "scoping" innovation epitomizes the regulations' approach to better EISs. That is, agencies are instructed to use "the scoping process, not only to identify significant environmental issues deserving of study, but also to deemphasize insignificant issues, narrowing the scope of the [EIS] process accordingly" (§1500.4(g)).

These objectives of CEQ's 1978 regulations have been bypassed in the 1980s. Agencies' first widespread shift in EIS writing seized on an authorization in section 1500.4(m), 14 places down in the regulations' implicit order of importance, which allowed agencies to circulate a FEIS consisting of a list of addenda to the DEIS when the changes were deemed "minor" (plus, of course, the EIS's public and interagency review record). A FEIS that consists of several pages of changes, referring to scattered pages in the text of a separate document, is extremely uninviting reading--especially for the reviewing decisionmakers and publics for whom the NEPA process is to be made useful. However, such a document is inexpensive to produce.

The Reagan administration's CEQ has concentrated its attention on refining guidance to agencies regarding categorical exclusions. CEQ's (1978) regulations (at §1500.4(p), 17 places down in the implicit order of importance) allow agencies to specify categories of projects or actions that are assumed per se to not have a significant environmental impact and thus are exempt from the regulations' EIS requirements. To be sure, many EISs were written during the 1970s on projects that were not major federal actions significantly affecting the environment. We drew at least two such projects among our sample of 29--the Weyerhaueser road across 100 feet of national forest land and the Skippanon River bridge replacing a span with the same allignment, location, and size. Nonetheless, categorical exclusion guidelines focus on fewer EISs, not better EISs (or better decisions).

Probably the primary flaw in CEQ's 1978 regulations is that they did not sufficiently force agencies to write better, shorter EISs. A 150-page limit leaves considerable space for extraneous information. A limit of about 75 pages—linked to strong substantive language mandating as, if not more, thorough analysis as has been produced to date—might really weed out extraneous verbiage. Projects of truly extraordinary complexity could be allowed 200 pages, with CEQ rationing variances for perhaps 25 such EISs per year. CEQ also needs to enforce its page limits, whether 75 or 150 pages. The Air Force's 1981 FEIS on "multiple protective shelter" basing of the MX missile totalled 58 volumes, CEQ's 300-page limit notwithstanding. CEQ ought to consider the model of the National Science Foundation, which imposed a 15-page limit on research proposals. The limit works because NSF enforced the rule by returning at least one major proposal to a principal investigator as "inappropriate for review because the proposal exceeds this limit," and publicized this rejection.[11]

In short, the empirical record and real-world limitations of the NEPA process present a grim prognosis for the rational, comprehensive, optimizing, scientific model of the prescriptive literature on EISs. As an alternative, we prefer CEQ's (1978) version of the old practical maxim: Keep It Simple and Succinct.

NOTES

1. Source: EPA's January 1983 index of EISs. The numbers cited include the numbers of final EISs, plus DEISs for which no FEISs had been filed, for the three agencies combined. Numbers of EISs only partially reflect the numbers of new projects; for example, so-called grandfathered projects (like Paint Creek Dam) and operation and maintenance EISs are included in the water project agencies' totals. As noted in Chapter 2, some agencies' EIS production increased in the late 1970s; the BLM's output rose from a 1971-1978 average of 9.8 per year to 40 per year during 1979-1982.

2. These documents, while not technically a random sample of all EISs, nonetheless strike us as a representative sample of mid-1970s EISs in terms of their distribution among the range of EIS-writing agencies, types of projects,

geographic locations, and proposal controversiality. Because of the intrinsic demands of our postproject audit design, the sample contains no EISs written under CEQ's (1978) formal NEPA regulations. As described in Chapter 2, EIS writing had become standardized by 1978, and we firmly believe that post-1978 EIS satisfy the rationalist ideal no better than pre-1978 documents.

3. These 115 cases consist of the time-ordered, experimental, and quantified data sets in Table 5.4 rated "exact" or "adequate," plus the "exact"-rated nominal measures (which could be regarded as good binary data). Also see Table 5.1.

4. Specifically, on a 1-5 grading-style scale in which "3" was defined as average quality, with "1" as "poor" and "5" as "excellent," All the documents averaged a 2.9 rating, U.S. EISs averaged 2.4, and the overall average for prediction quality was 2.8 (Rosenberg et al. 1981: 605, 612).

5. That is, because county-level predictions overestimate as often as they underestimate, 52% versus 48%, the authors averaged the absolute values of prediction errors.

6. Personal communication, Orie Loucks, May 26, 1987. Loucks argues that NEPA, as a beginning step in bringing about rational-scientific analysis of resource decisions, should be viewed positively, especially in comparison with the dearth of systematic analysis before the rationalist movement of the 1960s. The vindication of that movement, he argues, is on the horizon.

7. This monitoring is mandated by 36 C.F.R. 219.12(k), the Forest Service's (1982) land management planning regulations.

8. Quoting NEPA sections 2 and 102(2)(G). Sacred texts can, of course, be quoted selectively. NEPA sections 2, 101, and 102 also refer to "the health and welfare of man," "the profound influences of population growth, high density urbanization, [and] industrial expansion," "the social, economic, and other requirements of present and future generations," "safe, healthful, productive, and esthetically and culturally pleasing surroundings," "historic [and] cultural . . . aspects of our natural heritage," and "the integrated use of the natural and social sciences"--all of which, CEQ's (1978: §1508.6) regulations notwithstanding, can be read as justification for giving social impacts first-class status in the EIS process.

9. The term "tiering," based on the CEQ (1978: §1508.28) regulations, refers to doing environmental analyses on components of a broad action on which an earlier EIS had been written. "Compartments" are the smallest manage-

ment units within ranger districts and "prescriptions" are specific management actions. Both terms evolved from narrow meanings in the context of timber management plans into broader multiple-use meanings by the 1980s. As an example of "tiering" on the district, ranger Thurmond noted that his staff's compartment no. 181 management plan is about three times longer than the original Hiwassee unit plan.

10. That is, policies highlighted in sections 1500.4, 1500.2(b), and 1500.2(e) are cross-referenced in the regulations with specific or substantive provisions in the body of the regulations. For example, the requirement of section 1502.8 that EISs be written in plain language is repeated in the introduction's subsection 1500.4(d). Section 1500.5 also highlights important changes in the NEPA process, as distinct from improvements in EIS writing per se.

The regulations also require public involvement in the NEPA process (§1500.2(d), §1506.6). We have consistently supported this external reform view of the NEPA process (Friesema and Culhane 1976, Culhane 1978, and Culhane, Armentano, and Friesema 1985). The regulations (e.g., §1503) do not, however, impose a sufficient duty on EIS writers to give good faith consideration to comments by EPA, Interior, and other agencies or individuals with expertise in environmental review.

11. According to Northwestern University's in-house newsletter, Research Notes (June 15, 1986, p. 1). Of course, it is unclear that CEQ could impose such draconian, quasisubstantive rules under the guise of interpretative procedural rules. The Air Force MX FEIS set contains 58 volumes with the same cover design and EIS designation; technically, only seven volumes constituted the official FEIS subject to CEQ's length limitation.

Appendix: Research Methods

This appendix summarizes the content analysis, field-work, and accuracy rating methods used in the study. These methods are more thoroughly described in five draft appendices, which are available as a separate report, Forecasts and Environmental Decisionmaking: Methodological Appendices (July 1987, 32 pp.), by mail from the authors at the Center for Urban Affairs and Policy Research, Northwestern University, Evanston, Illinois 60208, or on file at the Northwestern University Library's NEPA Collection.

SAMPLE SELECTION METHODS

The selection of the implementation study sample, delineated in Table 3.1, began with a computer-generated random drawing from the EISs catalogued in the Environmental Protection Agency's (1980) index of EISs. Of the 400 entries initially selected, 249 were excluded from the sample for the reasons explained below. Another five cases were deleted because their implementation status was unobtainable.

Seven "non-cases" in the random sample are attributable to selection flaws in which the EIS-identification random number corresponded to a blank entry in our source or a duplicate case. Since our investigation would focus solely on the contents of final EISs, 111 draft EISs and supplements to draft statements were excluded. Draft EISs are working documents, subject to change before an agency commits itself to a project and the anticipated impacts. Agencies can reasonably be held accountable only for final impact predictions, not for predictions subject to revision.

In general, documents filed before 1973 (75 EISs) or after 1977 (36 EISs) were excluded from the sample for two reasons. Very early EISs would be distorted by the learning process that characterized the early history of NEPA. Very late EISs would have a substantially lower probability of project completion prior to the fieldwork and would leave too short a post-implementation period for data collection and evaluation. However, so as not to restrict the sample too severely or ignore occasional vagaries of the EIS filing process, we allowed very late 1972 and very early 1978 EISs to remain in the implementation study sample.

Field research constraints and our desire to avoid receiving a "Golden Fleece Award" necessitated the exclusion of 11 projects. Although the EIS for wind farms on the beaches of Hawaii was particularly appealing, a limited field research budget made it prudent to confine our field research to projects in the lower continental United States. The last set of deletions consisted of nine generic impact statements pertaining to multistate regions, the nation as a whole, or foreign countries, since the broad scope of predicted impacts in such statements would be extremely difficult to audit.

Two data collection methods were employed to determine the implementation status of the remaining 151 projects. Initially a telephone interview was designed to establish the nature of the project and its implementation status, to ascertain whether or not any delays or modifications had occurred, and to identify the reasons for implementation difficulties. Telephone contacts provided data for 146 projects. In addition, a questionnaire about project implementation was returned for 95 of these cases, or 63% of the initial sample of 151. The mailed survey served to confirm and clarify data already obtained, while also providing an opportunity for written comment by the contacted agencies.

Our survey data were supplied by government agency or business firm officials responsible for or knowledgeable about the sample projects. For some projects, more than one informant was identified. Most were regional or state administrators for the lead federal agencies, with a few notable exceptions. The Washington, D.C., office of the Nuclear Regulatory Commission, for example, supplied information for all the NRC EISs. At the other extreme, local offices of the Soil Conservation Service and the National Park Service were the typical contacts for those agencies. For the highways, state agencies were often most informative. In a few community development projects sponsored by the Department of Housing and Urban Development, private-sector contacts were able to provide more detailed data on project implementation than government sources.

For five EISs written by assorted agencies, however, neither form of data collection provided adequate information on implementation status. These involved two highway projects, an urban renewal project, a water resource project, and an annex to a military base. We attribute missing information mainly to an absence of institutional memory within the contacted office, coupled with the inability of that office to point us in the right direction for an alternative information source. However, we could not ascertain the implementation status of the military base proposal because of the sensitive national security nature of the project. The problems in obtaining status data on the five projects precluded their continued inclusion in both the implementation study and field research samples. After excluding projects for the above reasons, the implementation study sample consisted of 146 EISs.

CONTENT ANALYSIS METHODS

A "forecast," the coding unit in this content analysis, is defined as an assertion in an EIS about how the proposed action will or will not affect some physical, biological, social, or economic condition during or following implementation. Descriptions of the proposed action were not coded as impacts, though statements about effects that would occur at the same time as the action were coded as "incidental impact" forecasts. Each forecast refers to a single impact type or unit of measurement. Forecasts' lengths varied from as short as a clause to a paragraph or more. The entire EIS, from the summary sheet through the responses to comments, was searched for forecasts; however, substantially similar forecasts in different parts of an EIS were treated as a single forecast. As examples, 58 of these forecasts are summarized in Chapter 6.

The content analysis consisted of the following six steps. First, a draft coding protocol was pretested on a sample of 12 EISs that were not involved in any other aspect of the study. Second, two members of the research team identified forecasts in the 29 field-sample EISs and transcribed these forecasts onto word-processing files. Third, the draft protocol was revised, tested, and rerevised through four versions to refine coding instructions. Fourth, two coders independently coded all forecasts.

Fifth, these two coders combined their codes on a master coding sheet, noting any instances of substantive disagreement on any item. Sixth, the principal investigator resolved coder disagreements as a tie-breaker. This process follows the recommended content analysis procedures (cf. Krippendorf 1980) by using multiple coders and an iteratively refined protocol, but deviates from recommended practices in some respects, such as the use of tie-breaking.

Copies of the 13-page final version of the protocol may be obtained from the authors. The protocol consisted of the following 12 coding items plus their principal codes:

1. Forecast Type: Mitigation measure, project objective, or impact (including codes for standard and "incidental" impacts).

2. Impact Category: Physiographic, biological, social, or economic. The protocol discouraged mixed codes (e.g., "physiographic-social"), and 98% of all forecasts were coded within the four primary categories.

3. Impact Type: The resource or environmental condition that was the subject of the forecast; 190 types in the final protocol.

4. Unit of Measurement: The scientific or technical measurement units of a predicted impact ; 69 measurement-unit codes in the final protocol (e.g., the modal code, "0," representing "no clear unit of measurement"). See Table 4.4.

5. Forecast Quantification: Quantified (including codes for time-series, single-number, multiple-indicator, bounded-values, and percentage), nonnumeric "no impact," and unquantified verbal forecasts.

6. Significance: (a) "Insignificant," (b) "moderately significant," and (c) "very significant" key words or synonyms, plus codes for forecasts that are quantified without an explicit statement of significance or contain no indication of significance.

7. Directionality: Numerical increase, decrease, or no change in a clear unit of measurement; plus direction not specifiable because of unclear measurement unit, dichotomous or nominal effect, explicitly forecast continued trends, and no change.

8. Certainty: Possibility implied by key words like "may" or "could," probability implied by key words like "will" or "will not," plus codes for quantified-probability forecasts, and cases in which

Table A.1
Coder disagreements, by coding item. Rates based on 12 items and 1,105 forecasts for 13,260 codes in the content analysis (C.A.) phase, 9 items and 288 forecasts for 2,592 codes in the accuracy summary (A.S.) phase, and a combined total of 15,852 codes. The "salience," "direction," and "significance" items were not coded in the "A.S." round.

	Rate		
Forecast Characteristic	C.A.	A.S.	Combined
Salience of impact	10.2%	n/a	n/a
Certainty of forecast	2.4%	28.8%	7.9%
Impact type	6.9%	10.1%	7.5%
Impact category	5.8%	2.1%	5.0%
Quantification	2.7%	12.5%	4.7%
Adverseness/beneficiality	1.6%	15.3%	4.5%
Population affected	2.0%	12.5%	4.2%
Time frame of impact	1.5%	11.1%	3.5%
Units of measurement	0.7%	9.4%	2.5%
Forecast type, impact/mit./obj.	2.2%	2.1%	2.2%
Direction (of change in measure)	2.4%	n/a	n/a
Imputed significance	1.8%	n/a	n/a
Totals	3.4%	11.5%	4.7%

an impact was either conditional on an intervening event or a logically obvious consequence of the proposed action.
9. Beneficiality: beneficial, neutral, or adverse.
10. Referent Population: The location of the resource or community affected, most commonly (a) the people or natural resources within or immediately adjacent to the project site or (b) residents or resources within the municipality or county near or surrounding the project.
11. Time Frame: Impacts limited to the construction or implementation period, short-term impacts that occur soon after construction but then return to the preproject equilibrium, long-term or permanent impacts, and intermittent or random impacts, such as

accidents.

12. Salience: This item--coded simply "High,"
"Moderate," "Minor," and "Trivial" after several
attempts to improve its coding--involved substantial
subjective judgment and apparent reliability
problems.

The reliability of this content analysis is nominally
good. The recorded intercoder disagreement rate is a low
3.36% (Table A.1). However, during the analysis of forecast
accuracy reported in Chapters 6 and 7, the principal
investigator independently recoded nine characteristics of
the 288 forecasts for which field data were gathered.
During this round of coding, 299 additional disagreements
were identified, for a disconcerting 11.5% rate of disagreement between his codes and the codes assigned during the
first content-analysis round. The second round of coding is
not comparable to the full content analysis. The field-
sample forecasts were generally controversial, EIS writers
tend to hedge on controversial impacts, and hedged passages
invite intercoder disagreement. For this and other reasons,
second-round coding decisions were not reconciled with or
substituted for first-round codes. The results presented in
the book are based solely on coding of the full set of 1,105
forecasts. Given the peculiarities of the two rounds of
coding, the true reliability rate for the full sample of
1,105 forecasts is probably in the 90-95% range.

The second-round disagreement rate casts doubt on the
reliability of individual codes, but our analysis of intercoder disagreement patterns suggests that there is little
systematic bias in the aggregated distribution reported in
Chapter 4. For the most part, disagreements canceled each
other out. When disagreements were not offsetting, our
analysis suggests that coders tended to use unusual or low-
frequency codes less often than was appropriate. The coding
of impact salience was the least reliable in the content
analysis, so this item receives minimal attention in
Chapter 4. The principal bias of the coding appears to be
that short-term impacts (on the time-frame indicator) may
have been undercoded in up to 21 forecasts.

FIELDWORK METHODS

The field research team consisted of five people, the
authors plus two research assistants. Their disciplinary
backgrounds include four Ph.D. degrees in political science,

a Ph.D. in economics, a J.D. (law), plus master's or bachelor's degrees in forestry, public administration, economics, and political science. Generally, a two-person team composed of an author and a research assistant took responsibility for a given case.

The teams conducted fieldwork from April through July of 1983. The 29 field-sample projects are located in ten geographic clusters. Two field teams spent about a week in each cluster's region. The field teams sought time-series data for 1970-1982 from existing public or private records. When data proved available from the late 1960s, on a monthly, quarterly, or other time-ordered basis, or in a non-time-series quantified form, they were taken in that form. Potential sources of data located outside the cluster areas were investigated by telephone during follow-up weeks between field trips.

While in the field, the research teams conducted 120 interviews with 37 federal lead-agency officials; 15 officials of governmental project sponsors (mostly state highway department oficials); 19 local, 14 state, and 6 federal government nonsponsor (mostly regulatory) officials; eight representatives of project-sponsor businesses and ten other local businessmen; plus five environmentalists, three local residents, two academics, and one newspaper editor. The primary purpose of the interviews was to identify unanticipated impacts, but most interview time was devoted to obtaining informants' observations about impacts or patterns in the data.

IMPACT DATA VALIDITY

Depending on a mix of factors, we rated the intrinsic validity and reliability of each forecast impact's data. Since the ratings reflect several factors, the three data-aptness codes represent a judgmental ordinal scale:

"Exact"--Data or information that unambiguously establishes the postproject state of a phenomenon, and in the same units of measurement and affected population as used in the forecast (97 cases);

"Adequate"--Data whose types, units of measurement, etc., are not perfect, but good enough to support a solid conclusion about the accuracy of a forecast (94 cases); and

"Low adequate"--Despite reliability or validity flaws, sufficient data to form a conclusion about the

279

Table A.2

Secondary or multiple types of data or information. Primary sources most definitively classify forecast accuracy. Secondary and tertiary data present contrasting or reinforcing evidence about forecast impacts, not simply additional indicators of the same type as the primary source. The 22 cases with three distinct types of data are a subset of the 101 cases with two types of data.

| Nature of Data | Source Level | | | |
	Primary	Secondary	Tertiary	Total
Time series	81	14	3	98
Other time-ordered data	8	1	0	9
Experimental or controlled study results	5	2	0	7
Preproject, postproject quantified data	13	4	0	17
Other numeric data	19	12	1	32
Nominal status information	44	10	1	55
Documentary evidence	5	4	2	11
Verbal/interview information	64	54	15	133
Totals	239	101	22	362

impact with minimal confidence (48 cases).

Data that were fatally flawed and interview information deemed unreliable were rated "inadequate" and excluded from further analysis (see rows 6 and 7 of Table 5.1).

For primary data, we preferred quantified data over verbal information, time-series or controlled-study results over other numeric data, data on the same population and in the same units of measurement as those of the forecast, and interview information corroborated by multiple interview sources or reliable documents, plus secondary multiple indicators, comparison data series, and good interview information. Descriptions of these data characteristics, supplementing those covered in Chapter 5, follow.

The distribution of the secondary and tertiary field data are shown in Table A.2 and compared with the distri-

Table A.3
Validity of data collected during fieldwork. The secondary
validity and population codes do not necessarily pertain to
parallel secondary data types.

Validity Classification	Primary Data	Secondary Classif'n
Measure of the actual phenomenon, or same indicator used in forecast.	89	12
Surrogate measure, good validity	16	3
Surrogate measure, questionable validity	12	5
Comparison/control group data acquired	--	36
Known data flaws	3	5
Small number of data points	12	14
Interview/verbal information:		
Multiple sources	51	32
Statements adverse to interest	2	1
Single source	35	33
Documentary, photographic evidence	11	16
Observation of site	8	11
Totals	239	168

bution of primary data.

Table A.3 shows the basic validity classification of the
field data. The first classification describes data that
either clearly cover the impact or are the same indicator as
used in the forecast. Some of the data collected represent
surrogates for indicators of the actual impact, with better
surrogates distinguished from worse ones. The last five
ratings apply to nonquantitative information. Information
obtained from multiple interviewees is treated as more
reliable than single-source information, as are admissions
by interviewees of behavior adverse to their interests. In
contrast to Table A.2, the secondary validity codes do not
necessarily pertain to a parallel data type on the coding
sheet. There are 128 secondary validity codes and 40
tertiary validity codes, versus 101 secondary and 22
tertiary data types. The primary validity and population
codes, however, represent the principal characterizations of
the primary data type.

The population of the field data is an important factor in rating a case's data aptness. Most primary data (195 cases, which logically include almost all the interview data) deal with the same populations as those of the forecast. If a forecast's population area was a large enough part of the area from which the data were drawn that project impacts should be observable within the data variance, the data population was rated "close" (31 cases). Populations rated "not close enough" (12 cases) did not seem to meet this criterion, and their data-aptness rating was "low adequate." A significant difference between a data population and a forecast's population was sufficient reason for classifying the data inadequate. The single case among the 239 with a "different" population involves an unanticipated impact, and the case's overall data quality rating is "better." (Six cases received a data quality rating of "better" because their data were more precise than the relevant forecasts. These cases were combined with the cases rated "exact" in the analyses reported in Chapter 5.) Secondary population codes were assigned to 65 data sets: 44 "same," 11 "close," 8 "not close enough," and 2 "different."

THE ACCURACY INDEX AND DATA RELIABILITY

In assigning accuracy index scores used in Chapter 7, we gave cases the ordinal numbers corresponding to their ratings' places in the rank order shown in Table 7.2, for example, "4" for a "close" rating, "3" for an "impact within the range of a vague forecast" rating, "2" for "exceeds" or "less" ratings, and "1" for "inconsistent" ratings. A case's initial ordinal number could be increased or decreased if secondary accuracy classifications differed significantly from the primary match classification. A case received an increment of +1 if a secondary match rating was two ranks higher than the primary rating, unless the secondary match classification also received a beneficiality code of "trivial," or if two secondary match ratings were one rank higher. A case received a decrement of -1 if (a) a non-"trivial" secondary match rating was two ranks lower than the primary rating, (b) two secondary ratings were one rank lower than the primary rating, (c) a secondary match was "inconsistent" or "un/underanticipated-adverse," or (d) the primary impact-direction code was "wrong" or "bad." However, because the burden of proof should favor accurate ratings, (e) the coder could waive the decrement if

Table A.4
Accuracy ratings' insensitivity to impact data quality. The data-aptness ratings and primary-data-types distributions are shown in Table 5.5. The rating-reliability flag code is described in the text of the appendix. The accuracy means are averages of the ordinal ranks shown in Table 7.2.

Data Characteristic	$\overline{X}$ Accuracy	N
Primary Data Type		
Experimental results	3.60	5
Time-series or time-ordered data	2.69	89
Preproject, postproject quantified data	2.46	13
Other quantified data	3.00	19
Nominal status information	3.18	44
Documentary evidence	2.80	5
Verbal/interview information	2.70	64
Multiple Audit Data		
One type of audit data only	2.80	138
Two types of audit data/information	2.84	101
Three or more types of data/information	2.95	22
Data Aptness		
Exact	2.86	97
Adequate	2.79	94
Low	2.75	48
Special Reliability Code Indicating Low Coder Confidence in Rating	2.76	47

the case's data quality was coded "low" or the special reliability code described below was flagged "questionable" and, further, (f) any decrement to the lowest ordinal, "1," could be overruled in the coder's judgment. Ten increments and 21 decrements were assigned among the 239 cases.

The accuracy index provides a simple indicator of the reliability of the field data on project impacts. As shown in Table A.4, cases audited based on quantified data had an accuracy average of 2.75, while cases classified based on verbal information averaged 2.70, both of which approximate the nonmitigation cases' averages in the sample. Cases whose data are rated as more apt measures of their forecast impacts are only slightly more accurate than cases with less valid data. In addition, while assigning accuracy classifications, the principal investigator used a special code to flag cases in which he felt low confidence in the accuracy classification. (The codes do not simply duplicate the "low" data aptness rating; data for 18 of the flagged cases were rated "adequate" or "exact.") These cases were as accurate as average. Most differences in data-quality codes are well within the normal ± 0.10 range discussed in Chapter 7. A few larger differences are influenced by the prevalence of mitigations in subsamples (e.g., "nominal" data types), as with most accuracy differences reported in Chapter 7, or the quirks of subsamples with few cases (e.g., the interestingly higher accuracy of cases with three or more data types). In other words, the audit's wide range of data quality apparently introduced no biases into our conclusions about forecast accuracy.

If no evidence points to reliability or bias problems, the study's accuracy ranking scheme does have one quirk bearing on the validity of our ordinal index. In dealing with inaccurate cases (i.e., those below the sample mean of 2.8), the index seems to be a valid summary measure. However, analyses of relatively accurate forecasts involve the 3.3-2.9 range on the distribution of average index scores. Forecasts can receive scores above the sample average with either "4" for an accurate "close" rating or "3" for an "impact within the range of a vague forecast" rating. Depending on one's point of view, the latter may be as distressing as an unambiguously inaccurate forecast.

Abbreviations

ADT	Averave daily traffic
AEC	Atomic Energy Commission
AUM	Animal-unit month
B/C	Benefit/cost analysis
BARC	Beltsville Agricultural Research Center
BLM	Bureau of Land Management
BOR	Bureau of Outdoor Recreation
CBD	Central business district
CEQ	Council on Environmental Quality
CFR	Code of Federal Regulations
DEIS	Draft environmental impact statement
DNR	Department of Natural Resources
DOT	Department of Transportation (state or federal)
EA	Environmental assessment
EDA	Economic Development Administration
EIS	Environmental impact statement
ELR	Environmental Law Reporter
EPA	Environmental Protection Agency
EO	Executive Order
ERDA	Energy Research and Development Administration
F.2d	Federal Reporter, Second Series (West)
F.Supp.	Federal Supplement (West)
FAA	Federal Aviation Administration
FEIS	Final environmental impact statement
FHwA	Federal Highway Administration
FONSI	Finding of no significant impact
HUD	Housing and Urban Development, Department of
ICC	Interstate Commerce Commission
IDoC	Illinois Department of Conservation
JTU	Jackson turbidity units
LWBR	Light-water breeder reactor

MBF	Thousand board feet
MX	Missile experimental (also "Peacekeeper")
NEPA	National Environmental Policy Act
NPS	National Park Service
NRC	Nuclear Regulatory Commission
NRDC	Natural Resources Defense Council
OMB	Office of Management and Budget
ORNL	Oak Ridge National Laboratory
PANE	People Against Nuclear Energy
PILT	Payments in lieu of (local property) taxes
PPB	Planning, programming, and budgeting (systems)
REA	Rural Electrification Administration
RMP	Resource management plan
SCRAP	Students Challenging Regulatory Agency Procedures
SCS	Soil Conservation Service
TIE	The Institute of Ecology
ZBB	Zero-based budgeting

Bibliography

BOOKS AND ARTICLES

Allison, Graham. 1970. Essence of Decision. Boston: Little, Brown.

Anderson, Charles. 1979. "The Place of Principles in Policy Analysis." American Political Science Review 73 (September): 711-723.

Anderson, Frederick. 1973. NEPA in the Courts. Baltimore: Johns Hopkins University Press.

Andrews, Richard. 1976. Environmental Policy and Administrative Change: Implementation of the National Environmental Policy Act. Lexington, Mass.: D. C. Heath.

Baker, Martin, Joseph Kaming, and Richard Morrison. 1977. Environmental Impact Statements: A Guide to Preparation and Review. New York: Practicing Law Institute.

Bartlett, Robert. 1986. "Rationality and the Logic of the National Environmental Policy Act." Environmental Professional 8: 105-111.

Bisset, Ronald. 1984. "Post-Development Audits to Investigate the Accuracy of Environmental Impact Predictions." Zeitschrift für Umweltpolitik 7 (April): 463-484.

Brewer, Garry, and Peter deLeon. 1983. The Foundation of Policy Analysis. Homewood, Ill.: Dorsey Press.

Burchell, Robert, and David Listokin. 1975. The Environmental Impact Handbook. New Brunswick, N.J.: Center for Urban Policy Research, Rutgers University.

Caldwell, Lynton. 1982. Science and the National Environmental Policy Act: Redirecting Policy Through Procedural Reform. University, Ala.: University of Alabama Press.

Campbell, Donald. 1969. "Reforms as Experiments." _American Psychologist_ 24 (April): 409-429.
-----, and Julian Stanley. 1963. _Experimental and Quasi-Experimental Designs for Research._ Chicago: Rand McNally.
Canter, Larry. 1977. _Environmental Impact Assessment._ New York: McGraw-Hill.
Carson, Rachel. 1962. _Silent Spring._ Greenwich, Conn.: Fawcett.
Chapman, Stephen. 1986. "Should We Restore Regulation Over the Airlines?" _Chicago Tribune_ (September 7) §3: 5.
Chase, Alston. 1986. _Playing God in Yellowstone: The Destruction of America's First National Park._ Boston: Atlantic Monthly Press.
Cheremisinoff, Paul, and Angelo Morresi. 1977. _Environmental Assessment and Impact Statement Handbook._ Ann Arbor, Mich.: Ann Arbor Science.
Clarke, Jeanne Nienaber, and Daniel McCool. 1985. _Staking Out the Terrain: Power Differentials Among Natural Resources Management Agencies._ Albany: State University of New York Press.
Clawson, Marion. 1981. _New Deal Planning._ Baltimore: Johns Hopkins University Press.
Cohen, Michael, James March, and Johan Olson. 1972. "A Garbage Can Model of Organizational Choice." _Administrative Science Quarterly_ 17 (1): 1-25.
Commoner, Barry. 1971. _The Closing Circle._ New York: Alfred Knopf.
Cook, Thomas, and Donald Campbell. 1979. _Quasi-Experimentation: Design and Analysis Issues for Field Settings._ Boston: Houghton Mifflin.
Cortner, Hanna. 1976. "A Case Analysis of Policy Implementation: The National Environmental Policy Act of 1969." _Natural Resources Journal_ 16 (April) 323-338.
Culhane, Paul. 1974. "Federal Agency Organizational Change in Response to Environmentalism." _Humboldt Journal of Social Relations_ 2 (Fall/Winter) 31-44.
-----. 1978. "Natural Resources Policy: Procedural Change and Substantive Environmentalism," In _Nationalizing Government,_ eds., Theodore Lowi and Alan Stone. Beverly Hills, Calif.: Sage Publications, pp. 201-262.
-----. 1981. _Public Lands Politics: Interest Group Influence on the Forest Service and Bureau of Land Management._ Baltimore: Johns Hopkins University.
-----. 1984. "Sagebrush Rebels in Office." In _Environmental Policy in the 1980s,_ eds., Norman Vig and Michael Kraft. Washington, D.C.: CQ Press, pp. 293-317.

-----. 1987. "Heading 'Em Off at the Pass: MX and the Public Lands Subgovernment." In Federal Lands Policy, ed., Phillip Foss. Westport, Conn.: Greenwood Press, pp. 91-110.

-----, and H. Paul Friesema. 1978. "Environmental Impact Statements and Research on Public Policy and Applied Sciences." In EIS Annual Review, Vol. 1, ed., Ned Cronin. Washington D.C.: Information Resources Press, pp. 18-40.

-----, and H. Paul Friesema. 1979. "Land Use Planning for the Public Lands." Natural Resources Journal 19 (January): 43-74.

-----, Thomas Armentano, and H. Paul Friesema. 1985. "State-of-the-Art Science and Environmental Assessments: The Case of Acid Deposition." Environmental Management 9 (September): 365-378.

Cyert, Richard, and James March. 1963. A Behavioral Theory of the Firm. Englewood Cliffs, N.J.: Prentice-Hall.

Davis, Otto, M. A. H. Dempster, and Aaron Wildavsky. 1965. "A Theory of the Budgetary Process." American Political Science Review 60 (September): 529-547.

DeSouza, Glen R. 1979. System Methods for Socioeconomic and Environmental Impact Analysis. Lexington, Mass.: Lexington Books.

Drucker, Peter. 1976. "A Users' Guide to MBO." Public Administration Review 36 (January): 12-19.

Eckstein, Otto. 1958. Water Resources Development: The Economics of Project Evaluation. Cambridge, Mass.: Harvard University Press.

Environmental Law Institute. 1980. "Enforcing the Commitments Made in Environmental Impact Statements." Environmental Law Reporter 10 (August): 10153-10158.

Fairfax, Sally. 1978. "A Disaster in the Environmental Movement." Science 199 (February 17): 743-748.

-----, and Barbara Andrews. 1979. "Debate Within and Debate Without: NEPA and the Redefinition of the 'Prudent Man' Rule." Natural Resources Journal 19 (July): 505-536.

Freudenburg, William. 1986. "Social Impact Assessment." Annual Review of Sociology, 1986 12: 451-478.

Friesema, H. Paul, and Paul Culhane. 1976. "Social Impacts, Politics, and the Environmental Impact Statement Process." Natural Resources Journal 16 (April): 339-356.

-----, James Caporaso, Robert Lineberry, et al. 1979. Aftermath: Communities After Natural Disasters. Beverly Hills, Calif.: Sage Publications.

Goldsmith, Richard, and William Banks. 1983. "Environmental Values: Institutional Responsibility and the Supreme Court." Harvard Environmental Law Review 7 (Winter): 1-40.

Greeno, J. Ladd. 1983. Environmental Auditing. New York: John Wiley.

Harrison, L. Lee. 1984. Environmental Auditing Handbook: A Guide to Corporate and Environmental Risk Management. New York: McGraw-Hill.

Hart, Stuart, and Gordon Enk. 1980. Green Goals and Greenbacks: State-Level Environmental Review Programs and their Associated Costs. Boulder, Colo.: Westveiw Press.

Kaufman, Herbert. 1960. The Forest Ranger: A Study in Administrative Behavior. Baltimore: Johns Hopkins University Press.

Krippendorff, Klaus. 1980. Content Analysis. Beverly Hills, Calif.: Sage Publications.

Leistritz, F. Larry, and Steven Murdock. 1981. The Socioeconomic Impact of Resource Development: Methods for Assessment. Boulder, Colo.: Westview Press.

Lindblom, Charles. 1959. "The Science of 'Muddling Through.'" Public Administration Review 19 (Spring): 78-88.

------. 1965. The Intelligence of Democracy. New York: Free Press.

Lineberry, Robert. 1977. American Public Policy. New York: Harper & Row.

Liroff, Richard. 1976. A National Policy for the Environment: NEPA and Its Aftermath. Bloomington: Indiana University Press.

Lustick, Ian. 1980. "Explaining the Variable Utility of Disjointed Incrementalism," American Political Science Review 74 (June): 342-353.

March, James, and Herbert Simon, 1958. Organizations. New York: John Wiley.

Mazmanian, Daniel, and Jeanne Nienaber. 1979. Can Organizations Change? Environmental Protection, Citizen Participation, and the Corps of Engineers. Washington, D.C.: Brookings Institution.

McCallum, David. 1987. "Follow-up to Environmental Impact Assessment: Learning from the Canadian Experience." Environmental Monitoring and Assessment 8 (May) 199-215.

McCleary, Richard, and Richard Hay. 1980. Applied Time Series Analysis for the Social Sciences. Beverly Hills, Calif.: Sage Publications.

Medford, Derek. 1973. Environmental Harassment or Technology Assessment? New York: Elsevier.

Messing, Marc, H. Paul Friesema, and David Morell. 1979. Centralized Power. Boston: Oelischliger, Gunn & Hain.

Murdock, Steve, F. Larry Leistritz, Rita Hamm, et al. 1982. "An Assessment of Socioeconomic Assessments: Utility, Accuracy, and Policy Considerations." Environmental Impact Assessment Review 3 (December): 333-350.

Nienaber, Jeanne, and Aaron Wildavsky. 1973. The Budgeting and Evaluation of Federal Recreation Programs. New York: Basic Books.

Novick, David, ed. 1965. Program Budgeting. Cambridge, Mass.: Harvard University Press.

Olson, Michael. 1984. "Conflict and Complexity: Goal Diversity and Organizational Search Effectiveness." American Political Science Review 78 (June): 435-451.

Orloff, Neil. 1978. The Environmental Impact Statement Process: A Guide to Citizen Action. Washington, D.C.: Information Resources Press.

Padgett, John. 1980. "Bounded Rationality in Budgetary Research." American Political Science Review 74 (June): 354-372.

Pressman, Jeffrey, and Aaron Wildavsky. 1973. Implementation. Berkeley: University of California Press.

Quade, E. S. 1966. "Systems Analysis Techniques for Public Policy-making." Santa Monica, Calif.: Rand Corporation. Reprinted in Perspectives on Public Bureaucracy, ed., Fred Kramer. Boston: Little, Brown, 1981, pp. 192-210.

Reed, Phillip. 1983. "Environmental Audits and Confidentiality." Environmental Law Reporter 10 (October): 10303-10308.

Reich, Charles. 1962. "Bureaucracy and the Forests." Santa Barbara, Calif., Center for the Study of Democratic Institutions, Occasional Paper. Reprinted in Politics and Economic Policy-Making, ed., James Anderson. Reading, Mass.: Addison-Wesley, 1970, pp. 414-434.

Rosenberg, D. M., V. H. Resh, S. S. Balling, et al. 1981. "Recent Trends in Environmental Impact Assessment." Canadian Journal of Fisheries and Aquatic Sciences 38 (5): 591-624.

Simon, Herbert. 1947. Administrative Behavior. New York: Macmillan.

-----. 1960. The New Science of Management Decision. New York: Harper & Row.

Stanfield, Rochelle. 1985. "Enough and Clean Enough?" National Journal 17 (August 17): 1876-1887.

Stimson, James. 1985. "Regression in Time and Space: A Statistical Essay." American Journal of Political Science 29 (November): 914-947.

Taylor, Serge. 1984. Making Bureaucracies Think: The Environmental Impact Statement Strategy of Adminis-trative Reform. Stanford, Calif.: Stanford University Press.

Thompson, Mark. 1980. Benefit-Cost Analysis for Program Evaluation. Beverly Hills, Calif.: Sage Publications.

von Neuman, John, and Oskar Morgenstern. 1944. Theory of Games and Economic Behavior. Princeton, N.J.: Princeton University Press.

Wandesford-Smith, Geoffrey. 1979. "Environmental Impact Assessment in the European Community." Zeitschrift fur Umweltpolitik 2 (1): 35-76.

Wenner, Lettie. 1982. The Environmental Decade in Court. Bloomington: Indiana University Press.

Wichelman, Allan. 1976. "Administrative Agency Implementation of the National Environmental Policy Act of 1969." Natural Resources Journal 16 (April): 263-300.

Wildavsky, Aaron. 1964. The Politics of the Budgetary Process. Boston: Little, Brown.

Williams, Chuck. 1982. "The Park Rebellion." Not Man Apart (June).

INSTITUTE OR GOVERNMENT REPORTS
AND UNPUBLISHED PAPERS

Andrews, Richard, Paul Cromwell, Gordon Enk, et al. 1977. Substantive Guidance for Environmental Impact Assessment. The Institute of Ecology.

Arthur D. Little, Inc. 1981. Environmental Auditing: A Forum on Current Corporate Practices and Future Trends. Cambridge, Mass.: A. D. Little, Inc., Center for Environmental Assurance, June.

Beanlands, Gordon, and Peter Duinker. 1983. An Ecological Framework for Environmental Impact Assessment in Canada. Halifax, Nova Scotia: Dalhousie University, Institute for Resource and Environmental Studies, in cooperation with Federal Environmental Assessment Review Office.

Chapman, P.M., G. A. Vigers, M. A. Farrell, et al. 1982. Survey of Biological Effects of Toxicants upon Puget Sound Biota: Broad-scale Toxicity Survey. Boulder, Colo.: NOAA Office of Marine Pollution, December.

Claggett, Lawrence. 1977. Fish Population Changes in an Unstocked Warmwater Reservoir During the Initial Two

Years Following Impoundment. Master's Thesis.
Columbus: Ohio State University.
Clark, Roger, George Stankey, and John Hendee. 1974. An Introduction to CODINVOLVE. Portland, Ore.: U.S. Forest Service, Pacific Northwest Forest and Range Experiment Station, #PNW-223, April.
Council on Environmental Quality. 1985. Environmental Quality--1984. Washington, D.C.: Government Printing Office.
Culhane, Paul, and H. Paul Friesema. 1977. "Why Environmental Assessments 'Fail'." Paper presented at the Annual Meeting of the American Society for Public Administration, Altanta, April.
----- and H. Paul Friesema. 1982. "A Study of the Precision of Environmental Impact Assessment," Proposal to the National Science Foundation. Evanston, Ill., Northwestern University, February.
-----, H. Paul Friesema, and Janice Beecher. 1985. Forecasts and Environmental Decisionmaking: Executive Summary, Report to the National Science Foundation. Evanston, Ill.: Center for Urban Affairs and Policy Research, Northwestern University, April.
Dames & Moore, Inc. 1983. "Commencement Bay Study Environmental Impact Assessment (COBS II) for U.S. Army Corps of Engineers, Seattle District." Dames & Moore, April.
Dean, James, and James Stimson. 1980. "Pooling Cross Sections and Time Series." Paper presented at the Annual Meeting of the Midwest Political Science Association, Chicago, April.
Dexter, R.N., D. E. Anderson, E. A. Quinlan, et al. 1981. A Summary of Knowledge of Puget Sound Related to Chemical Contaminants. Boulder, Colo.: NOAA Office of Marine Pollution Assessment, December.
Duinker, Peter. 1985. "Forecasting Environmental Impacts: Better Quantitative and Wrong Than Qualitative and Untestable." Paper presented at the Audit of Environmental Assessment Results Conference, Banff, Alberta, October.
Enk, Gordon, and Stuart Hart. 1980. Improving the Quality and Utility of Scientific and Technical Information in Environmental Impact Statements. Report of a Symposium for State and Federal Decision Makers. Rensselaerville, N.Y., Institute on Man and Science, June.
Environmental Law Institute. 1981. NEPA in Action: Environmental Offices in Nineteen Federal Agencies. Washington, D.C.: Government Printing Office, October.
Environmental Protection Agency. 1980. "Cumulative Index

of Environmental Impact Statements, Arranged by Agency and Bureau, as of May 28, 1980." Washington, D.C.: Environmental Protection Agency, Office of Federal Activities [computer printout].

Environmental Resources Ltd. 1984. Prediction in Environmental Impact Assessment. Gravenhage, Netherlands: Ministerie van Volkhuisvesting, Ruimtelijke Ordening en Milieubeheer, March.

Friesema, H. Paul. 1975. "Environmental Group Fragmentation and Administrative Decision-Making." Paper at the Annual Meeting of the Midwest Political Science Association, Chicago.

------. 1981. The Scientific Content of Environmental Impact Statements: Workshop Conclusions. Report to Indiana University Center for Advanced Studies in Science, Technology, and Public Policy. Indianapolis: The Institute of Ecology, May.

Hunsaker, Donald, and Donald Lee. 1985. "Assessing the Environmental Impacts of Low Probability, High Consequence Events: A Follow-up Study." Paper presented at the Audit of Environmental Assessment Results Conference, Banff, Alberta, October.

Institute of Ecology, University of Georgia. 1971. Optimum Pathway Matrtix Analysis Approach to the Environmental Decision Making Process. Athens, Ga.: University of Georgia, Institute of Ecology.

Johnson, Art, Bill Yake, and Dale Norton. 1983. "A Summary of Priority Pollutant Data for Point Sources and Sediment in Inner Commencement Bay: Part 3, Blair Waterway." Olympia, Wash.: Washington State Department of Ecology, July.

Leopold, Luna, Frank Clarke, Bruce Hanshaw, et al. 1971. A Procedure for Evaluating Environmental Impact. Geological Survey Circular 645. Washington, D.C.: Government Printing Office.

Loucks, Orie. 1982. "Evaluating and Predicting Ecological Effects of Coal Combustion for Electric Power: An Overview of the Wisconsin Power Plant Study." Indianapolis, Ind., and Madison, Wisc.: The Institute of Ecology, January.

Malins, Donald, B. B. McCain, D. W. Brown, et al. 1982. Chemical Contaminants and Abnormalities in Fish and Invertibrates from Puget Sound. Boulder, Colo.: NOAA Office of Marine Pollution Assessment, June.

Marcus, Linda. 1979. A Methodology for Post-EIS Monitoring. Geological Survey Circular 782. Washington, D.C.: Government Printing Office.

Munro, David, Thomas Bryant, and A. Matte-Baker. 1986. Learning from Experience: A State-of-the-Art Review and Evaluation of Environmental Impact Assessment Audits. Hull, Quebec: Canadian Environmental Assessment Research Council.

National Park Service. 1982. Resource Protection Case Study: Grand Teton National Park, Jackson Hole. Denver, Colo.: Department of the Interior, June.

Ruckel, H. Anthony. 1980. "Annual Report FY 1979-80, Rocky Mountain Office, Sierra Club Legal Defense Fund." Denver, Colo.: Sierra Club Legal Defense Fund, September 10.

------. 1985. "Annual Report on the Activities of the Rocky Mountain Office of SCLDF, FY 1984-1985." Denver, Colo.: Sierra Club Legal Defense Fund, August 30.

Russell, Geoff. 1982. "Benthic Macroinvertebrate Fauna of the Robert S. Kerr Reservoir Below Lake Tenkiller Adjacent to the Effluent Outfall of the Sequoyah UF_6 Facility, Kerr-McGee Nuclear Corporation, Gore, Oklahoma, October 1980 to December 1981." Stillwater, Okla.: Oklahoma State University, July.

Warner, Maurice, and Edward Preston. 1973. A Review of Environmental Impact Assessment Methodologies. Washington, D.C.: Environmental Protection Agency, October.

Weil, Ray, and Frank Payer. 1983. Phosphorus Renovation of Wastewater by Overland Flow Land Application. College Park, Md.: Maryland Water Resources Research Center, August.

Winder, John, and Ruth Allen. 1975. The Environmental Impact Assessment Project: A Critical Appraisal. Washington, D.C.: The Institute of Ecology, October.

ENVIRONMENTAL IMPACT STATEMENTS

Agricultural Research Service. 1974. Proposed Modernization and Related Expansion of Two Sewage Wastewater Treatment Plants for the Beltsville Agricultural Research Center, Final Environmental Impact Statement. Beltsville, Md.: Agricultural Research Service, October.

Atomic Energy Commission. 1972. Monticello Nuclear Generating Plant, Final Environmental Statement. Washington, D.C.: Atomic Energy Commission, November.

------. 1974. Radioactive Waste Facilities, Oak Ridge National Laboratory, Environmental Impact. Washington, D.C.: Atomic Energy Commisssion, August.

Bureau of Outdoor Recreation. 1973. Proposed Illinois Beach Acquisition, Lake County, Illinois, Final Environmental Statement. Washington, D.C.: Bureau of Outdoor Recreation, January.

Corps of Engineers. 1971. La Farge Lake, Kickapoo River, Vernon County, Wisconsin, Draft Environmental Statement. St. Paul, Minn.: Corps of Engineers, September.

─────. 1973. Diked Disposal Area Site No. 12, Cleveland Harbor, Ohio, Final Environmental Impact Statement. Buffalo, N.Y.: Corps of Engineers, January.

─────. 1974a. Weymouth-Fore and Town Rivers, Boston Harbor, Rock Removal, Final Environmental Statement. Waltham, Mass.: Corps of Engineers, April.

─────. 1974b. Paint Creek Lake, Final Environmental Statement. Huntington, W. Va.: Corps of Engineers, July.

─────. 1975. Maintenence Dredging, Blair Waterway, Repair and Extension of East Training Wall, Puyallup River, Tacoma Harbor, Final Environmental Impact Statement. Seattle, Wash.: Corps of Engineers, December.

─────. 1977. Cross-Florida Barge Canal Restudy Report, Final Environmental Impact Statement. Jacksonville, Fla.: Corps of Engineers, February.

Geological Survey. 1975. Proposed Plan of Mining and Reclamation: Belle Ayr South Mine, Amax Coal Company, Campbell County, Wyoming, Draft Environmental Statement. Washington, D.C.: Department of the Interior, Geological Survey, March.

Department of Housing and Urban Development. 1977a. Residential Development of Clovis Heights and Cougar Estates, Clovis, California, Final Environmental Impact Statement. San Francisco: Department of Housing and Urban Development, March.

─────. 1977b. Newington Forest Project, Final Environmental Impact Statement. Washington, D.C.: Department of Housing and Urban Development, September.

─────. 1977c. Shepard Park West, Shepard Park Development, St. Paul, Minnesota, Final Environmental Impact Statement. Minneapolis: Department of Housing and Urban Development, November.

Economic Development Administration. 1976. Yakima Central Business District Improvements, City of Yakima, Washington, Final Environmental Impact Statement. Washington, D.C.: Economic Development Administration, July.

Energy Research and Development Administration. 1976. Light Water Breeder Reactor Program, Final Environmental Statement. Washington, D.C.: Energy Research and Development Administration, 5 volumes, June.

Environmental Protection Agency. 1978. Henrico County, Virginia, Wastewater Treatment Facilities, Final Environmental Impact Statement. Philadelphia: Environmental Protection Agency, March.

Federal Aviation Administration. 1974. Establishment of an Air Route Surveillance Radar Facility, Cummington, Massachusetts, Final Environmental Impact Statement. Burlington, Mass.: Federal Aviation Administration, July.

Florida Department of Transportation. 1973. Upgrading of State Road 24 between I-75 and North-South Drive to a Six-Lane Facility, the Provision of a SR 24 - I-75 Interchange, and the Improvement and Extension of SR 728 Between SR 24 and SR 331 as a Four-Lane Facility, Final Environmental Statement. Tallahassee: Florida DOT and Federal Highway Administration, August.

------. 1974. Construction of Interstate Route 295 from Old Kings Road to West of Interstate Route 95 North of Jacksonville, Florida, Final Environmental Statement. Tallahassee: Florida DOT and Federal Highway Administration, February.

Forest Service. 1974a. Weyerhaueser Company Road Construction Proposal in Hanson Creek Share Cost Agreement Area, King County, Washington, Final Environmental Statement. Portland, Ore.: Forest Service, June.

------. 1974b. Mineral King Recreation Development, Draft Environmental Statement. San Francisco: Forest Service, December.

------. 1975. Hiwassee Unit, Final Environmental Statement. Cleveland, Tenn.: Forest Service, June.

------. 1976a. Mineral King, Final Environmental Statement. San Francisco: Forest Service, February.

------. 1976b. Ozone Unit Plan, Final Environmental Statement. Russellville, Ark.: Forest Service, August.

Michigan Department of State Highways. 1973. I-94 - Lake Shore Drive Interchange Reconstruction, Final Environmental Statement. Lansing: Michigan D.S.H. and Federal Highway Administration, September.

National Park Service. 1974. Actions Under Consideration, Jackson Hole Airport, Grand Teton National Park, Wyoming, Final Environmental Statement. Omaha, Neb.: National Park Service, March.

------. 1975. Master Plan, Grand Teton National Park, Wyoming, Final Environmental Statement. Omaha, Neb.: National Park Service, September.

Nuclear Regulatory Commission. 1975. Sequoyah Uranium Hexaflouride Plant, Final Environmental Statement.

Washington, D.C.: Nuclear Regulatory Commission, February.

Oregon State Highway Division. 1975. Skippanon River Bridge, Warrenton, Oregon, Final Environmental Impact Statement. Salem: Oregon S.H.D. and Federal Highway Administration, December.

Rural Electrification Administration. 1975. Alma Unit No. 6 and Related 161 kV Transmission Lines, Final Environmental Impact Statement. Washington, D.C.: Rural Electrification Administration, 6 volumes, May.

Soil Conservation Service. 1976. South Fourche Watershed, Final Environmental Impact Statement. Little Rock, Ark.: Soil Conservation Service, March.

Wisconsin Department of Transportation. 1973. Connorsville - East County Line Road, S.T.H. 64, Dunn County, Final Environmental Statement. Madison: Wisconsin DOT and Federal Highway Administration, October.

------. 1975. USH 53 and USH 8, Barron and Chippewa Counties, Final Environmental Impact Statement. Madison: Wisconsin DOT and Federal Highway Administration, August.

LEGAL CITATIONS

Administrative Procedures Act of 1946. 5 U.S. Code 500-526.

National Historic Preservation Act of 1966. 16 U.S. Code 470 et seq.

Barta v. Brinegar, 358 F.Supp. 1025 (W.D. Wisc., May 1973).

Calvert Cliffs' Coordinating Committee v. Atomic Energy Commission, 449 F.2d 1109 (D.C. Cir., July 1971).

Catholic Action of Hawaii v. Brown, 468 F.Supp. 190 (D. Hawaii, March 1979), 623 F.2d 602 (9th Cir., July 1980); Weinberger v. Catholic Action of Hawaii, 454 U.S. 139 (December 1981).

Citizens to Preserve Overton Park v. Volpe, 432 F.2d 1307 (6th Cir., September 1970), 401 U.S. 402 (March 1971).

City of Rochester v. U.S. Postal Service, 541 F.2d 967 (2nd Cir., September 1976).

Committee for Nuclear Responsibility v. Seaborg 462 F.2d 783 (D.C. Cir., October 1971); Committee for Nuclear Responsibility v. Schlesinger, 404 U.S. 917 (November 1971).

Council on Environmental Quality. 1970. "Statements on Proposed Federal Actions Affecting the Environment: Interim Guidelines." 35 Federal Register 7390-7393 (May 12).

-----. 1971. "Guidelines for Statements on Proposed Federal Actions Affecting the Environment." 36 Federal Register 7724-7729 (April 23).

-----. 1973. "Preparation of Environmental Impact Statements: Guidelines." 38 Federal Register 20550-20562 (August 1).

-----. 1978. "National Environmental Policy Act: Interpretation of Procedural Provisions, Final Regulations." 40 C.F.R. pts. 1500-1508; 43 Federal Register 55978-56007 (November 29).

Cummington Preservation Committee v. Federal Aviation Administration, 524 F2d 241 (1st Cir., October 1975).

Environmental Defense Fund v. Corps of Engineers, 324 F.Supp. 878 (D. D.C., January 1971); In Re Cross-Florida Barge Canal, 329 F.Supp. 543 (M.D. Fla., July 1971); Canal Authority of Florida v. Callaway, 4 ELR 20259 (M.D. Fla., January 1974), 489 F.2d 567 (February 1974), 512 F.2d 670 (5th Cir., May 1975).

Environmental Defense Fund v. Corps of Engineers, [Gillham dam] 325 F.Supp. 728,749 (E.D. Ark., February 1971), 342 F.Supp. 1211 (E.D. Ark., May 1972), 470 F.2d 289 (8th Cir., November 1972).

Executive Order 11,514. "Protection and Enhancement of Environmental Quality." 3 C.F.R. 286; 35 Federal Register 4247 (March 5, 1970).

Executive Order 11,593. "Protection and Enhancement of the Cultural Environment." 16 U.S. Code 470 (Supp. 1, 1971); 36 Federal Register 8921 (May 15, 1971).

Forest Service. 1982. "National Forest System Land and Resource Management Planning." 47 Federal Register 43026-43052 (September 30).

Hanly v. Mitchell 460 F.2d 640 (May 1972); Hanly v. Kliendienst, 3 ELR 20016 (S.D. N.Y., August 1972), 471 F.2d 823 (December 1972).

Hiram Clarke Civic Club v. Lynn, 476 F.2d 421 (5th Cir., April 1973).

Image of Greater San Antonio v. Brown, 570 F.2d 517 (5th Cir., March 1978).

Kleppe v. Sierra Club, 427 U.S. 390 (June 1976).

McDowell v. Schlesinger, 404 F.Supp 221 (W.D. Mo., July 1975).

National Environmental Policy Act of 1969. 42 U.S. Code 4321-4361. Public Law 91-190 (1970).

Natural Resources Defense Council v. Morton, [OCS leasing] 337 F.Supp. 165 (D. D.C., December 1971), 458 F.2d 827 (D.C. Cir., January 1972).

Natural Resources Defense Council v. NRC, 547 F.2d 633, and

Aeschliman v. NRC 547 F.2d 622 (D.C. Cir., July 1976); Vermont Yankee Nuclear Power Corp. v. Natural Resources Defense Council, 435 U.S. 519 (April 1978); 685 F.2d 459 (D.C. Cir., April 1982); Baltimore Gas and Electric Co. v. NRDC, 462 U.S. 87 (June 1983).

Northern States Power v. Minnesota, 320 F.Supp. 172 (D. Minn., December 1970), 447 F.2d 1143 (8th Cir., September 1971).

Oregon Environmental Council v. Kunzman, 12 ELR 20769 (D. Ore., May 1982), 714 F.2d 901 (9th Cir., August 1983), 614 F.Supp. 657 (D. Ore., April 1985).

People Against Nuclear Energy v. NRC, 678 F.2d 222 (D.C. Cir., May 1982); Metropolitan Edison Co. v. PANE, 460 U.S. 766 (April 1983).

Scenic Rivers Association of Oklahoma v. Lynn, 382 F.Supp. 69 (E.D. Okla., September 1974), 520 F.2d 240 (10th Cir., July 1975); Flint Ridge Dev. Co. v. Scenic Rivers Ass'n, 426 U.S. 776 (June 1976).

Scientists' Institute for Public Information v. AEC, 481 F.2d 1079 (June 1973).

Sierra Club v. Morton, [Northern Great Plains coal development] 421 F.Supp. 638 (D. D.C., February 1974), 514 F.2d 856 (D.C. Cir., June 1975); Kleppe v. Sierra Club, 427 U.S. 390 (June 1976).

Sierra Club v. Morton, [National Wildlife Refuge system] 395 F.Supp. 1187 (D. D.C., June 1975), 581 F.2d 895 (D.C. Cir., May 1978); Andrus v. Sierra Club, 442 U.S. 347 (June 1979).

Students Challenging Regulatory Agency Procedures v. U.S., [SCRAP I] 353 F.Supp. 317 (D. D.C., January 1973), 412 U.S. 669 (June 1973); [SCRAP II] 371 F.Supp. 1291 (D. D.C., February 1974); Aberdeen and Rockfish R.R. v. SCRAP, 422 U.S. 289 (June 1975).

Trinity Episcopal School Corp. v. Romney, 523 F.2d 88 (2nd Cir., July 1975); Karlen v. Harris, 590 F.2d 39 (2nd Cir., December 1978); Strycker's Bay Neighborhood Council v. Karlen, 444 U.S. 223 (January 1980).

Wilderness Society v. Hickel, [Alaska pipeline] 325 F.Supp 422 (D. D.C., April 1970); Wilderness Society v. Morton, 463 F.2d 1261 (D.C. Cir., May 1972), 479 F.2d 842 (D.C. Cir., February 1973); [attorneys fees awarded] 495 F.2d 1026 (D.C. Cir., April 1974); Alyeska Pipeline Co. v. Wilderness Society, 421 U.S. 240 (May 1975).

Index

301

Biological impacts, 12, 262–263
accuracy of, 234–235
data on, 128, 135–136
examples of, 157, 175, 189–190, 200, 214
forecasts about, 87–88, 104–107, 112–113
Bureau of Land Management, 30, 180, 270n
Bureau of Outoor Recreation. See Illinois Beach State Park expansion

Caldwell, Lynton, 6, 8, 9, 11, 14, 25, 257, 258
Campbell, Donald, 121, 139
Canter, Larry, 10–13, 22n, 83, 101, 113, 258
Cleveland harbor disposal site 12, 68, 78, 188
data on, 148, 188–189, 225n
impacts of, 188–192, 225n
Clovis Heights subdivision, 113–114, 266
Columbia station, study of, 120
Content analysis, 85, 95, 275–278
Controversiality, 125
and accuracy, 250
and forecast precision, 109–111, 118n
Corps of Engineers, 257
See also Cleveland harbor disposal site 12, Cross-Florida barge canal, Paint Creek dam, Tacoma Harbor channel maintenance, Weymouth-Fore and Town Rivers channel project
Council on Environmental Quality, 23–24, 26, 38, 40–41, 43, 216, 250, 272n
NEPA guidelines of, 13, 28, 40–43, 298–299

1978 regulations of, 10, 29, 35–38, 43, 95, 262, 267–270, 272n, 299
Cross-Florida barge canal
controversy over, 110–111, 118n, 250
description of, 69–70, 77–79
Cummington radar facility
description of, 70, 78
forecasts about, 104, 111, 118n

Decision theory, 2–6
Direction of impact, accuracy of, 140, 228–229
"Dutch study," 84, 112, 259

Economic Development Administration. See Yakima CBD project
Economic impacts, 8, 11, 12–13, 31–33, 41
accuracy of, 234–235
data on, 129, 134–135, 136
examples of, 157–162, 175–176, 186–188, 191–192, 193, 203–212, 219–221,
forecasts about, 88–90, 106–107
Energy, 88, 107–108, 129, 191–192, 205–208, 237–238
See also Alma unit no. 6, Monticello nuclear station, Oak Ridge National Laboratory waste facility, Sequoyah uranium hexafluoride plant, Shippingport breeder reactor
Energy Research and Development Administration. See Shippingport breeder reactor
Environmental assessments, 29, 37

Environmental impact
statements
draft EISs, 37, 42, 48,
269, 273
format of, 38-43
"human environment" in, 30-
33, 95, 262
legal requirements for, 15
need for, 29-30
writers of, 33-34, 50-54,
255n
Environmental Protection
Agency, 48, 54, 169, 191-
192, 270n, 272n
See also Henrico County
wastewater system
External reform model, 16-18,
108-111, 249-250, 256n,
272n

Federal Aviation Administration, See also Cummington
radar facility, Jackson
airport
Federal Highway Administration, 248
See also Highways
Field data, 126-131
biases in, 134-136
data aptness scale for, 132
validity of, 132-134, 279-
282
Field sample, 64-76
completion status of, 76-79
implementation pathologies
of, 77-79
selection of, 64-66, 273-
275
Fieldwork methods, 124-127,
278-279
Finding of no significant
impacts (FONSI), 37
Florida Department of Transportation. See State Road
24 interchange (Florida),
I-295 beltway (Jackson-
ville, Florida)

Forecast impact, definition
of, 142, 229
Forecasts, 97, 102
imprecision of, 104-111
impact likelihood, 99-101
of "no impact," 96, 102
quantification of, 96, 105-
106, 117n, 132-133
types of, 85-94, 106
verbal, 96-101, 117n
See also Vague forecasts
Forest Service, 181, 248-249,
261, 264
See also Hiwassee unit
plan, Mineral King
recreation development,
Ozone unit plan,
Weyerhaeuser road

Geological Survey, 103, 153-
155
Government buildings, See
Cummington radar
facility, Beltsville
Agricultural Research
Center sewage plants
Grand Teton National Park
master plan, 72, 178-179
data on, 148, 177-179
impacts of 177-188
See also Jackson airport

Henrico County wastewater
system, 115
description of, 72, 78
impacts of, 215
Highway 64 (Dunn County,
Wisconsin), 68, 78
Highways, See State Road 24
interchange (Florida),
I-94 interchange (Michi-
gan), I-295 beltway
(Jacksonville, Florida),
Skipanon River bridge,
U.S. 53 and U.S. 8
(Wisconsin), Highway 64
(Dunn County, Wisconsin)

303

304

description of, 71, 78-79
impacts of, 198-200, 265-266
Shippingport breeder reactor, 75-76, 78
controversy over, 110, 118n, 250
impacts of, 253
Sierra Club, 38, 74, 79, 109, 117n, 196
Simon, Herbert, 2-6, 16, 254
Skipanon River bridge
description of, 67-68, 78, 226n
impacts of, 217-219
Social impacts, 8, 11, 12-13, 31-33, 41, 262-263
accuracy of, 233-236
data on, 128-129, 134-135, 136
examples of, 162-165, 177, 179-182, 190-191, 193-200, 215, 217-219, 222-223,
forecasts about, 88-90, 94-96
Soil Conservation Service, 257
See also South Fourche small watershed project
South Fourche small watershed project, 69, 78
State Road 24 interchange (Florida), 66-67, 78

Tacoma Harbor channel maintenance, 68, 78, 226n
forecasts about, 103
impacts of, 216-217

Unanticipated impacts, 174, 214-217, 229-231, 253
Units of measurement
of field data, 130-131, 137
of forecasts, 99-100
Urban development. See

Henrico County wastewater system, Yakima CBD project
U.S. 53 and U.S. 8 (Wisconsin)
description of, 67, 77-78
impacts of, 196-198

Vague forecasts, 85, 112, 265-266
accuracy of, 253, 259
examples of, 148-152, 163-165, 171-173, 175-176, 177, 181, 189-190, 198-200
Vagueness index, 104-105, 109-110, 112

Water resources. See Cross-Florida barge canal re-study, Cleveland harbor disposal site 12, Paint Creek dam, South Fourche small watershed project, Tacoma Harbor channel maintenance, Weyerhaeuser road
Watt, James, 40, 74, 110, 196, 225n
Weyerhaeuser road, 73, 269
controversy over, 110
data on, 145
forecasts about, 107, 108, 110
impacts of, 221-222
Weymouth-Fore and Town Rivers channel project
description of, 69, 78
impacts of, 193-195, 219-221, 254n
Wisconsin Department of Transportation. See Highway 64 (Dunn County), U.S. 53 and U.S. 8

Yakima CBD project, 70-71, 78